LECTURE NOTES IN
DEEP LEARNING
Theoretical Insights into an Artificial Mind

LECTURE NOTES IN
DEEP LEARNING
Theoretical Insights into an Artificial Mind

Shlomo Dubnov
University of California, San Diego, USA

Dongmian Zou
Duke Kunshan University, China

W💡 World Scientific

NEW JERSEY · LONDON · SINGAPORE · BEIJING · SHANGHAI · HONG KONG · TAIPEI · CHENNAI · TOKYO

Published by

World Scientific Publishing Co. Pte. Ltd.

5 Toh Tuck Link, Singapore 596224

USA office: 27 Warren Street, Suite 401-402, Hackensack, NJ 07601

UK office: 57 Shelton Street, Covent Garden, London WC2H 9HE

Library of Congress Cataloging-in-Publication Data

Names: Dubnov, Shlomo author | Zou, Dongmian author
Title: Lecture notes in deep learning : theoretical insights into an artificial mind / Shlomo Dubnov,
    University of California, San Diego, USA, Dongmian Zou, Duke Kunshan University, China.
Description: New Jersey : World Scientific, [2026] | Includes index.
Identifiers: LCCN 2024056893 | ISBN 9789811280627 hardcover |
    ISBN 9789811281570 paperback | ISBN 9789811280634 ebook |
    ISBN 9789811280641 ebook other
Subjects: LCSH: Deep learning (Machine learning) | Neural networks (Computer science)
Classification: LCC Q325.73 .D83 2026 | DDC 006.3/1--dc23/eng/20250602
LC record available at https://lccn.loc.gov/2024056893

British Library Cataloguing-in-Publication Data
A catalogue record for this book is available from the British Library.

For any available supplementary material, please visit
https://www.worldscientific.com/worldscibooks/10.1142/13524#t=suppl

Desk Editors: Soundararajan Raghuraman/Steven Patt

Typeset by Stallion Press
Email: enquiries@stallionpress.com

To my dear family, wife Lior, our six kids and their families,
and our parents.

Shlomo Dubnov

...

To my sweet wife Jennie, my daughter Olivia, and my parents.

Dongmian Zou

Preface

Why the Book?

This book aims to provide a comprehensive introduction to deep learning, a vibrant subfield of machine learning that has quickly become a cornerstone of modern artificial intelligence. The motivation for writing this text originated when we co-taught a deep learning course for senior students at Duke Kunshan University (DKU). When we prepared for the course, we found many excellent textbooks and resources. These include classics like the *Deep Learning Bible* by Goodfellow *et al.* (2016) and the recent deep learning textbook by Bishop and Bishop (2023) developed upon Bishop's "Bible for machine learning". There are also more hands-on texts such as *Deep Learning with Python* by Ketkar and Santana (2017). However, we struggled to find a resource that precisely aligned with our specific educational goals.

Our students, already familiar with the fundamentals of machine learning, were particularly eager to explore how neural networks could offer a novel perspective of data analysis. We discovered that most existing texts were either too exhaustive, covering more breadth than depth, or shied away from the essential technical details that are crucial for a deep, functional understanding of neural network models. On the other hand, texts consisting of mainly theories are not suitable for early-entry students. This book is our attempt to address these shortcomings. It is structured to guide readers from fundamental principles of neural network modeling and optimization through to complex considerations of neural networks as Gaussian processes, and explores neural tangent and information theory.

We aimed to provide a broad introduction that covers the main mathematical concepts in deep learning in a level accessible to upper-level undergraduate engineering students. Advanced chapters were designed to offer an additional intuition into the field by linking deep learning to broader statistical and information modeling approaches.

In crafting this book, we placed a significant emphasis on conciseness and directness, mirroring the style often found in "lecture notes". This approach is intentional, reflecting our desire to deliver the most impactful content without overwhelming readers with overly complex explanations or derivations. We have distilled the essence of deep learning into digestible discussions. The streamlined presentation is particularly suited to senior undergraduate students and others who already possess a grounding in machine learning but need a direct route to the advanced concepts that are shaping the future of artificial intelligence.

Structure of the Book

This book contains six parts; each is roughly designed to progressively build upon the previous ones. It starts with basic concepts and leads up to more specialized discussions.

The first part of this book presents basics of neural networks. We begin with the simplest neural network architectures and quickly demonstrate practical implementations, which are useful for readers keen to apply these concepts to real-world data. This section also addresses why neural networks are particularly effective and discusses the optimization techniques uniquely suited to these models.

The second part focuses on autoencoders and variational autoencoders, which are fundamental tools for learning representations for data. This part connects these ideas back to classical techniques like PCA and probabilistic PCA, which provide a deeper understanding of data representation.

The third part deals with special neural network architectures that have been critical to the current advancements in AI.

These include convolutional neural networks (CNNs), recurrent neural networks (RNNs), and transformers. We explain their development and applications in vision, audio, and natural language processing.

The fourth part presents generative models. We expand upon variational autoencoders (VAEs) and then study generative adversarial networks (GANs). These are fundamental generative models which relate to very interesting mathematical and statistical theories. Further, we study normalizing flows and diffusion models, which are driving recent developments in AIGC (artificial intelligence generated content).

The fifth part introduces more theoretical part of deep learning. Rather than focusing on applications of neural networks, this part is dedicated to understanding the theories behind neural networks, situating them within the broader context of traditional learning theory.

The sixth part contains some further topics we believe to be of critical importance, including transfer learning, explainable AI, and deep reinforcement learning.

How To Use the Book

This book can be used as a textbook for a semester-long undergraduate course that assumes students' familiarity with machine learning, or a short course in deep learning. By the end of the course, students will be able to comprehend the fundamentals of deep learning; design, train, and implement neural network models; understand the mathematical, statistical, and computational issues of deep learning; touch the state-of-the-art solutions to practical applications.

The six parts of this book fit well both 7-week or 14-week semester courses. Taking a 14-week course as an example, we can spend two weeks on each part of the book, after a week's review of machine learning and a week's presentation.

If this book is used as the reference for a short course, it is possible to skip some technicalities, such as optimization methods and information theory. It is recommended to focus more on Part two to Part four, which cover many interesting applications that should keep beginners interested.

Acknowledgments

The development of this book has been significantly shaped by the feedback and contributions from our students. Their feedbacks have enriched the content, and we are deeply grateful for their involvement. We thank (in alphabetical order) Lihui Chen, Xinmeng Chen, Huangrui Chu, Zhiyue Feng, Shuyi Guan, Yike Guo, Wanqi Hu, Rui Jiang, Zilin Jiang, Alex Jin, Ran Ju, Yu Leng, Teodora Eliyanova Petkova, Eric Qu, Jiyang Tang, Qixuan Wang, Shouju Wang, Sissi Wang, Zepu Wang, Yijia Xue, and Chenglin Zhang. We also thank the Division of Natural and Applied Sciences of DKU for their support in maintaining small class sizes and allowing us the privilege of teaching this course over three consecutive years.

We hope this book serves not only as an educational tool but also as a source of inspiration for students and professionals alike, encouraging further exploration and innovation in the field of deep learning.

Contents

PART 1

Neural Network Basics

Chapter 1

Introduction

1.1 Deep Learning: What and Why

In 2016, during a five-game series of Go that garnered worldwide attention, a computer program defeated Lee Se-dol, one of the top human players at that time. Nowadays, despite the familiarity people in all occupations have with the concept of *artificial intelligence*, it is still hard to believe that the key ideas behind *deep learning*, the very backbone of this remarkable computer algorithm, have been developed several decades before that pivotal moment already.

That being said, a lot of people still find themselves unsure about the true essence of "deep learning". To those well versed in mathematics, the idea of deep learning is so similar to Kolmogorov's solution to Hilbert's 13th Problem, in which one takes superposition of simple functions to create more complex functions. Indeed, in the realm of deep learning, a *layer* is a building block of operations similar to a simple function, and a deep *neural network* can be created by cascading these layers. In order to adapt to one's task, a deep neural network needs to "learn" from data and determine all the details about its layers.

Even prior to the advent of deep learning, the concept of "learning" was not new in the field of statistics and computer science. For instance, a simple one-dimensional linear regressor also needs to learn from data in order to determine its slope and intercept. However, what sets deep learning apart from these "shallow" learning approaches is the intricate composition of its simple layers. This indicates that there are so many parameters in a deep learning model, which require a large amount of computing resources. Consequently, the popularity of deep learning has surged over the past decade, largely due to the recent advancements in graphic processing units (GPUs). It is widely believed that the invention of AlexNet (Kirzhevsky *et al.*, 2012) was a milestone in the brief history of deep learning. In essence, the multi-layer structure indeed is the reason behind the success of AlexNet. Starting from shallow layers, AlexNet starts to gather local features from the input images, which in a way constitute the higher-level features in deeper layers.

To the general public, the concept of "big data" is probably more familiar than "deep learning". In this context, "big" refers not only to the large amount of available datasets but also to the fact that those data live in high-dimensional spaces. Deep learning, however, has emerged as a powerful tool that leverages the potential of big data and continuously adapts to address the associated challenges.

We begin our journey by reviewing the key concepts in machine learning that serve as the foundation for deep learning. Subsequently, we introduce the main architecture employed throughout this book, oftentimes a synonym of deep learning: (artificial) neural networks.

1.2   From Linear Models to Neural Networks

In machine learning, our primary objective is to extract meaningful patterns from a dataset. The outcome of a machine learning algorithm is typically a *model*, which can be seen as a function designed to make predictions based on the input. For instance, in regression problems, the input of the model can be features of an instance, and the output can be the predicted value. In a more formal setting, this model takes input from a certain space, namely the domain set, and we hope that it not only fits the specific examples in our dataset but also generalizes effectively to the entire domain set. Of course, we don't want the search space for the model to be very large. Therefore, we typically constrain the search space to a "hypothesis class," which limits the types of models we consider to those that align with our pre-existing beliefs or theoretical preferences.

For instance, if we believe that the output of the model depends linearly on the input, we can construct a linear model. Say the input of the model is an n-dimensional vector (representing n different features of an instance) and the output is a scalar (representing the predicted value). We can represent our model in the form of

$$y \equiv f(\mathbf{x}) = \mathbf{w}^\top \mathbf{x} + b, \tag{1.1}$$

where $\mathbf{x} \in \mathbb{R}^n$ is the input and $y \equiv f(\mathbf{x}) \in \mathbb{R}$ is the output. Here, $\mathbf{w} \in \mathbb{R}^n$ and $b \in \mathbb{R}$ are *parameters* of the model, which we need to learn from the available examples in the dataset. We can refer to $\mathbf{w}$ as the *weights* of the model and b as the *bias* of the model.

Suppose we have a dataset $\mathbb{X} \subset \mathbb{R}^n$ containing N data points, known as the *training set*. We can write $\mathbb{X} = \{\mathbf{x}_i\}_{i=1}^N$, where each $\mathbf{x}_i \in \mathbb{R}^n$. A more convenient way is to represent $\mathbb{X}$ using a matrix $\mathbf{X} \in \mathbb{R}^{N \times n}$, where each row represents a data point. That is, the ith row of $\mathbf{X}$ is given by $\mathbf{x}_i^\top$, and $\mathbf{X}$ can be written as

$$\mathbf{X} = \begin{bmatrix} - \ \mathbf{x}_1^\top \ - \\ - \ \mathbf{x}_2^\top \ - \\ \vdots \\ - \ \mathbf{x}_N^\top \ - \end{bmatrix}.$$

This matrix representation aligns with the convention commonly used in many coding languages such as Python.

By utilizing this matrix representation, our model can generate predictions as a vector in $\mathbb{R}^N$ for all the data points at once, which can be conveniently summarized as

$$f(\mathbf{X}; \mathbf{w}, b) = \mathbf{X}\mathbf{w} + b\mathbf{1}, \tag{1.2}$$

where $\mathbf{1} = (1, \ldots, 1)^\top \in \mathbb{R}^N$ is the vector of all 1's.

The model in (1.2) may not be able to fit all the data points exactly due to the inherent bias we introduced by constraining the model to be linear. Nevertheless, this is not necessarily a problem. In regression (and other supervised learning tasks as well), our primary aim is to *generalize* what we learn from the dataset to unseen portions of the data distribution. It is acceptable for the model to capture the main trends while allowing for minor errors in the training data. Indeed, a perfect fit for the training data using a very complex model may indicate poor generalization to new data, known as *overfitting*. However, significant errors in fitting the training data may indicate that the model is *underfitting,* which could compromise its performance on new data.

A classical example is the following. Suppose that we need to learn from a dataset of four points in $\mathbb{R}^2$: $\mathbf{x}_1 = (0,0)^\top$, $\mathbf{x}_2 = (0,1)^\top$, $\mathbf{x}_3 = (1,0)^\top$, $\mathbf{x}_4 = (1,1)^\top$. The target responses are $y_1 = y_4 = 0$ and $y_2 = y_3 = 1$. We can consider this dataset as showing all values of the "exclusive OR" or "XOR" function which returns 1 if and only if exactly one of the two input arguments is 1. Figure 1.1 illustrates the dataset.

Clearly, a linear function cannot fit the dataset perfectly. This is because, due to linearity, the function value on the line segment connecting $\mathbf{x}_1$ and $\mathbf{x}_4$ would have to be constantly 0, and for the same reason, the function value on the line segment connecting $\mathbf{x}_2$ and $\mathbf{x}_3$ has to be constantly 1. However, the fact that these two line segments meet at $(0.5, 0.5)^\top$ immediately leads to a contradiction.

In order to resolve this problem, we must go beyond linear functions and incorporate nonlinearity. For instance, an example of a desired nonlinear function is illustrated in Figure 1.2. The height axis of the figure represents the function value, indicating the output of the nonlinear function.

A simple approach to constructing nonlinear functions is to apply a pointwise nonlinear *activation function* to the output of a linear transformation. That is, we can take

$$f(\mathbf{x}) = \sigma(\mathbf{w}^\top \mathbf{x} + b), \tag{1.3}$$

where σ is a function applied to each element of $\mathbf{w}^\top \mathbf{x} + b$. The output of the activation function is commonly referred to as an *activation.*

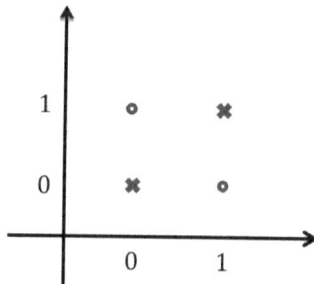

Fig. 1.1. An illustration of the "XOR" dataset. The coordinates of the points represent $\mathbf{x}_1, \ldots, \mathbf{x}_4$, respectively. Circles represent the points with a value of 0, and crosses represent the points with a value of 1.

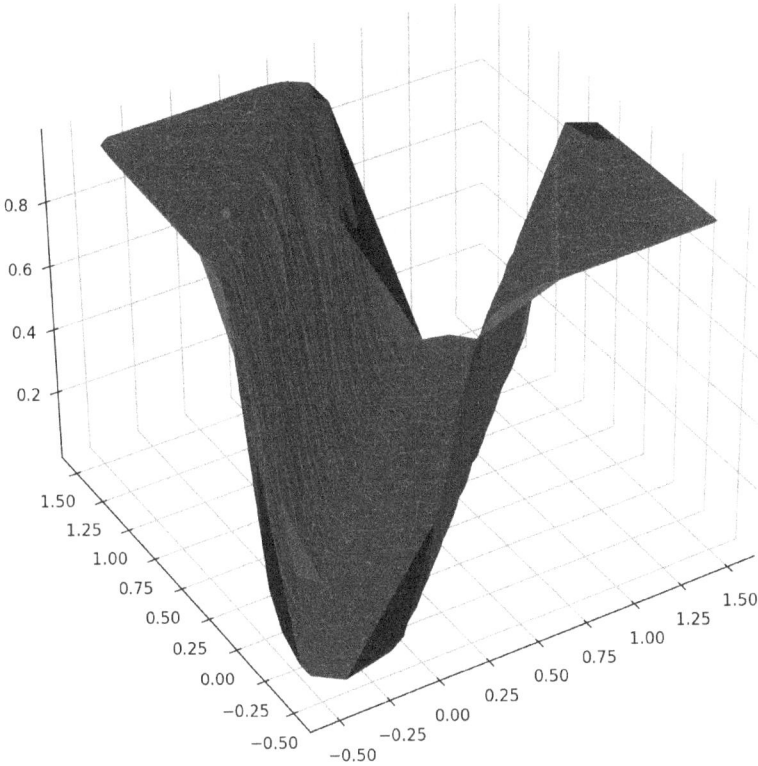

Fig. 1.2. A nonlinear function that perfectly fits the "XOR" dataset. This image is credited to Alex Jin.

More complex functions can be created by cascading multiple nonlinear functions like (1.3).

In the following, we discuss some popular choices of nonlinear functions used in neural networks:

- **Sign function:**

$$\text{Sign}(x) = \begin{cases} 1, & \text{if } x > 0 \\ 0, & \text{otherwise.} \end{cases} \tag{1.4}$$

- **Rectified linear unit (ReLU):**

$$\text{ReLU}(x) = \begin{cases} 0, & \text{if } x \leq 0 \\ x, & \text{otherwise.} \end{cases} \tag{1.5}$$

ReLU is a very simple activation function, which is efficient to calculate. The derivative is also very simple, which benefits the gradient-based learning algorithms. Moreover, a part of activations will be zero after applying the ReLU function, yielding sparsity of neurons. ReLU has shown great success in practice, especially in deep networks.

- **Hyperbolic tangent (tanh):**

$$\tanh(x) = \frac{e^x - e^{-x}}{e^x + e^{-x}}. \tag{1.6}$$

Another activation function with a similar shape is the sigmoid function, also known as the logistic function:

$$\text{sigmoid}(x) = \frac{1}{1 + e^{-x}}. \tag{1.7}$$

Figure 1.3 plots the tanh and sigmoid functions. We remark that in the literature "sigmoid" may be used to refer to any function of a similar "S" shape. tanh has been used more widely than σ since its output is zero-centered and thus the input of the next layer of the neural network is more balanced. It is especially used a lot in recurrent neural networks.

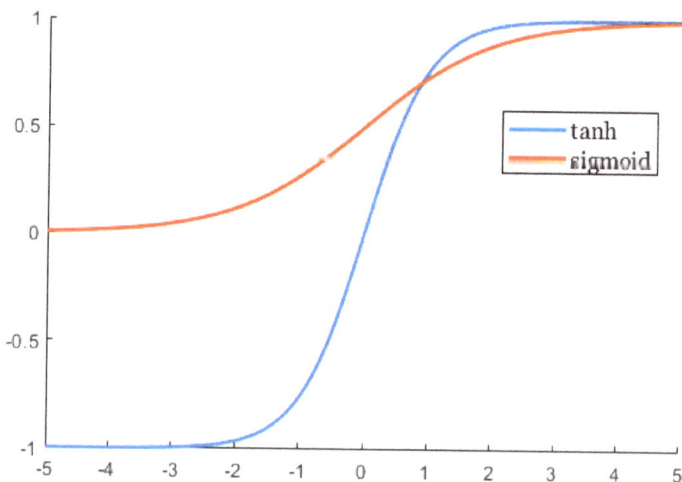

Fig. 1.3. Plots of the tanh and sigmoid functions.

- **Leaky ReLU:** given $0 < \alpha < 1$,

$$\text{LeakyReLU}(x) = \begin{cases} \alpha x, & \text{if } x \le 0 \\ x, & \text{otherwise.} \end{cases} \tag{1.8}$$

Here, α is usually taken to be a small positive number (e.g., $\alpha = 0.01$ is the default value in PyTorch). Unlike ReLU, when the input of a neuron is less than zero, Leaky ReLU keeps the neurons alive throughout the learning process. This means that the neurons can continue to learn.

- **GELU (Hendrycks and Gimpel, 2016):**

$$\text{GELU}(x) = \frac{1}{2}x \left(1 + \text{erf} \left(\frac{x}{\sqrt{2}} \right) \right), \tag{1.9}$$

where erf is the Gaussian error function defined by

$$\text{erf}(x) = \frac{2}{\sqrt{\pi}} \int_0^x e^{-t^2} dt. \tag{1.10}$$

GELU provides a smooth, non-monotonic function that approximates the ReLU activation function. It has gained popularity, particularly in the context of transformer models.

If we only use one linear transformation followed by one activation function, the *expressivity* of our model will still be limited. To this end, we may cascade a sequence of linear transformations and activation functions to create a multi-layer model. Let's discuss this in detail. Let $\mathbf{x} \in \mathbb{R}^n$ be the input. First, we can apply n_1 linear transformations, followed by an activation function σ, to obtain n_1 values; or equivalently, we apply a multi-dimensional linear transformation, followed by σ, to obtain a vector $\mathbf{h}_1 \in \mathbb{R}^{n_1}$, say $\mathbf{h}_1 = \sigma(\mathbf{W}_1^\top \mathbf{x} + \mathbf{b}_1)$, where $\mathbf{W}_1 \in \mathbb{R}^{n \times n_1}$ and $\mathbf{b}_1 \in \mathbb{R}^{n_1}$. This is our first layer. Next, in the second layer, we apply another linear transformation, followed by an activation function (taken to be the same σ for simplicity), to get $\mathbf{h}_2 = \sigma(\mathbf{W}_2^\top \mathbf{h}_1 + \mathbf{b}_2) \in \mathbb{R}^{n_2}$, where $\mathbf{W}_2 \in \mathbb{R}^{n_1 \times n_2}$ and $\mathbf{b}_2 \in \mathbb{R}^{n_2}$. Following this trend, in the lth layer, we have $\mathbf{h}_l = \sigma(\mathbf{W}_l^\top \mathbf{h}_{l-1} + \mathbf{b}_l)$, $l = 1, 2, \ldots$ (with the understanding that $\mathbf{h}_0 = \mathbf{x}$), where $\mathbf{W}_l \in \mathbb{R}^{n_{l-1} \times n_l}$ and $\mathbf{b} \in \mathbb{R}^{n_l}$. Suppose we have L layers in total, we may output $\mathbf{z} = \mathbf{h}_L \in \mathbb{R}^{n_L}$. If we want a scalar output, we simply take $n_L = 1$.

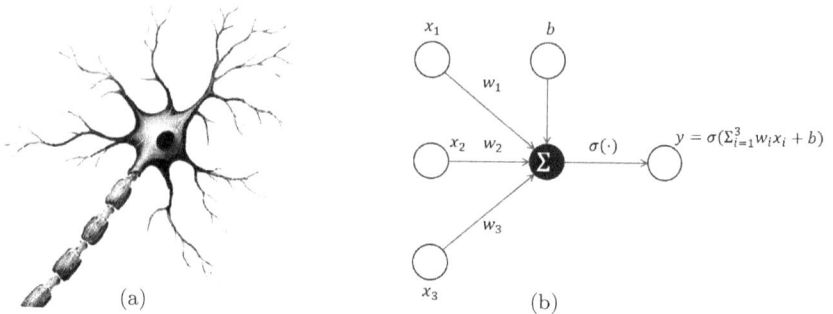

Fig. 1.4. A neural network is similar to how human brains work. (a) Structure of human neural network; (b) structure of artificial neural network.

The cascading we have in the above is called an artificial neural network (ANN), or a neural network (NN). It mimics how a human brain neural network works. Figure 1.4 illustrates a simple structure of a human neural network.

In the above cascading, the input $\mathbf{x}$ is called the *input layer* of the neural network. Instead of regarding it as an n-dimensional vector, we can imagine it as n neurons, each carrying a scalar value. The intermediate $\mathbf{h}_1, \ldots, \mathbf{h}_{L-1}$ are called *hidden layers*. The output $\mathbf{z} = \mathbf{h}_L$ is called the *output layer*.

A deep neural network refers to a neural network with multiple layers. The neural network constructed as above is also known as a *fully connected network* (FCN), a *dense* network, or a *multi-layer perceptron* (MLP). It is also custom to call each layer a fully connected layer or a dense layer.

1.3 Training Neural Networks

The training or learning process of a machine learning algorithm guides us in finding the best parameters with which the model captures the relationship between the input data and the corresponding target responses. For linear models, closed-form solutions exist and we can derive them in a single step. Another typical algorithm, which is more generic, starts with randomized weights and bias and then updates them iteratively. Consider the linear model (1.2) discussed in the previous section. Suppose we initialize the weights and bias to be $\mathbf{w}^{(0)}$ and $b^{(0)}$, respectively. Our initialized model will output the following results given the input $\mathbf{X}$:

$$f(\mathbf{X}; \mathbf{w}^{(0)}, b^{(0)}) = \mathbf{X}\mathbf{w}^{(0)} + b^{(0)}\mathbf{1}. \tag{1.11}$$

Since we have actually learned nothing yet from data, the performance of $f(\mathbf{X}; \mathbf{w}^{(0)}, b^{(0)})$ may not be good. Suppose $\mathbf{y} \in \mathbb{R}^N$ is the vector of target responses, or *labels*, available from our dataset. Then, the performance of our current model can be measured by comparing $\mathbf{y}$ with $f(\mathbf{X}; \mathbf{w}^{(0)}, b^{(0)})$. This performance should provide us with some guidance to update the parameters. In other words, we need to find $\mathbf{w}^{(1)}$ and $b^{(1)}$ so that the performance is improved after replacing $\mathbf{w}^{(0)}$ and $b^{(0)}$ with them. Afterwards, we may consecutively find a sequence of parameters $\mathbf{x}^{(t)}$ and $b^{(t)}$, $t = 0, 1, 2, \ldots$, so that the performance keeps improving. The process in which we update these parameters is called *training*.

The same principle holds for training a neural network. Specifically, we can initialize a random collection of parameters and then update them iteratively, guided by the model's performance on the dataset. The performance can be measured by a *loss* function $\ell = \ell(\boldsymbol{\theta}; \mathbf{x})$, where $\boldsymbol{\theta}$ denotes the collection of all parameters including $\mathbf{W}_l$ and $\mathbf{b}_l$, $l = 1, \ldots, L$. The semicolon splitting $\boldsymbol{\theta}$ and $\mathbf{x}$ emphasizes the fact that the error depends on both the model parameters and the input. Given an input $\mathbf{x}$ and its corresponding target response $\mathbf{y}$, by writing the output of the neural network as $f(\mathbf{x}, \boldsymbol{\theta})$, we can measure their difference using the squared norm $\|\mathbf{y} - f(\mathbf{x}, \boldsymbol{\theta})\|^2$ as an error, or a loss, for the single input $\mathbf{x}$. The smaller the error, the better the performance of our model for the input $\mathbf{x}$. Considering all the data points $\mathbb{X} = \{\mathbf{x}_1, \ldots, \mathbf{x}_N\}$ and the corresponding target responses $\mathbb{Y} = \{\mathbf{y}_1, \ldots, \mathbf{y}_N\}$, we can define the overall loss function, widely known as the mean squared error (MSE) loss

$$\ell(\boldsymbol{\theta}; \mathbb{X}) = \frac{1}{N} \sum_{i=1}^{N} \left[\|\mathbf{y}_i - f(\mathbf{x}_i, \boldsymbol{\theta})\|^2 \right]. \tag{1.12}$$

The most widely used optimization method for minimizing the overall loss function in neural networks is the *gradient descent* method. Given the overall loss function ℓ and the current parameters $\boldsymbol{\theta}^{(t)}$, the negation of the gradient of the loss function, namely $-\nabla_{\boldsymbol{\theta}}\ell(\boldsymbol{\theta}^{(t)})$, provides for the parameters a direction, along which the function value descends locally. That is, if the parameters move along this direction for a small step, the function value ℓ necessarily decreases. The step size is called the *learning rate*. Let's call the learning rate η, with $\eta > 0$. Then, we can update the parameters according to

$$\boldsymbol{\theta}^{(t+1)} = \boldsymbol{\theta}^{(t)} - \eta \nabla_{\boldsymbol{\theta}} \ell(\theta^{(t)}). \tag{1.13}$$

Starting with $\boldsymbol{\theta}^{(0)}$, the gradient descent method executes (1.13) for $t = 0, 1, \ldots$ until a prescribed number of steps is reached.

In the loss function (1.12), the mean operator $\frac{1}{N} \sum_{i=1}^{N} [\cdot]$ can be treated as a practical implementation of the *expectation*, taken over the data distribution $\mathcal{D}$. If we have sufficient information about $\mathcal{D}$ and a groundtruth labeling function $\mathbf{y}(\mathbf{x})$, then the MSE loss function should have been written as

$$\ell(\boldsymbol{\theta}) = \mathbb{E}_{\mathbf{x} \sim \mathcal{D}} \left[\|\mathbf{y}(\mathbf{x}) - f(\mathbf{x}, \boldsymbol{\theta})\|^2 \right]. \tag{1.14}$$

More generally, instead of the squared normed difference, we can consider a generic error function $\mathcal{E}(\mathbf{x}; \boldsymbol{\theta})$. The corresponding overall loss function is then

$$\ell(\boldsymbol{\theta}) = \mathbb{E}_{\mathbf{x} \sim \mathcal{D}} \left[\mathcal{E}(\mathbf{x}; \boldsymbol{\theta}) \right]. \tag{1.15}$$

By linearity of the expectation operator,

$$\nabla_{\boldsymbol{\theta}} \ell(\boldsymbol{\theta}) = \mathbb{E}_{\mathbf{x} \sim \mathcal{D}} \left[\nabla_{\boldsymbol{\theta}} \mathcal{E}(\mathbf{x}; \boldsymbol{\theta}) \right]. \tag{1.16}$$

Methods in which gradient steps are taken following (1.16) are generally called *stochastic gradient descent* (SGD) methods. In this perspective, the gradient $\nabla_{\boldsymbol{\theta}} \ell(\boldsymbol{\theta})$ in the gradient descent method can be computed as long as we have a good way of estimating $\mathbb{E}_{\mathbf{x} \sim \mathcal{D}}$. Previously, in each gradient step, we use the full batch of training data to estimate the gradient:

$$\mathbb{E}_{\mathbf{x} \sim \mathcal{D}}[\cdot] \approx \frac{1}{N} \sum_{i=1}^{N} [\cdot]. \tag{1.17}$$

However, when the dataset is too big, using the full batch can be computationally inefficient and memory demanding. There are two alternative approaches to approximating $\mathbb{E}_{\mathbf{x} \sim \mathcal{D}}$: the *online* approach and the *mini-batch* approach.

In the online approach, $\mathbb{E}_{\mathbf{x}\sim\mathcal{D}}[\nabla_{\boldsymbol{\theta}}\mathcal{E}(\mathbf{x};\boldsymbol{\theta})] \approx \nabla_{\boldsymbol{\theta}}\mathcal{E}(\mathbf{x}_i;\boldsymbol{\theta})$ for a single data point $\mathbf{x}_i$. This means that the gradient update is performed using one data point at a time. In the mini-batch approach,

$$\mathbb{E}_{\mathbf{x}\sim\mathcal{D}}[\nabla_{\boldsymbol{\theta}}\mathcal{E}(\mathbf{x};\boldsymbol{\theta})] \approx \frac{1}{|\mathbb{B}|}\sum_{i\in\mathbb{B}}\nabla_{\boldsymbol{\theta}}\mathcal{E}(\mathbf{x}_i;\boldsymbol{\theta}), \tag{1.18}$$

where $\mathbb{B}$ denotes the indices of a mini-batch, a small subset of the training dataset. To guarantee that we make good use of the full training data, instead of randomly drawing samples $\mathbb{B}$ at each step, we usually randomly partition the training data into mini-batches at the very beginning and feed them into our algorithm sequentially. This ensures that each mini-batch equally contributes to the gradient update. A complete pass through the entire training data is referred to as one *epoch*.

In practice, selecting an appropriate learning rate η is crucial for the success of the training process. A small η leads to slow convergence, which means that it takes too many iterations to reach the optimal solution. On the other hand, a large η may cause the value of the loss function to fluctuate or even increase. One common approach to choosing a good learning rate is through a process called *validation*. In this approach, a dataset called the validation set, independent of the training data, is used. The validation set is not used for updating the model parameters. Instead, it serves as a benchmark to compare the performance of the model trained with different learning rates.

The learning rate here is an example of a *hyperparameter*. We predetermine a set of learning rates and choose the best performing one according to the performance of our trained model on the validation set.

It is also possible to utilize a learning rate finder to automatically determine an appropriate learning rate. The process involves starting with a very small learning rate and gradually increasing it while monitoring the behavior of the loss function on the current mini-batch. The learning rate is increased as long as it is safe to do so (the loss does not worsen).

A more systematic way to adjust the learning rate is to use a *scheduler*. For instance, the 1-cycle scheduler (Smith and Topin, 2019) has a schedule that contains a warmup phase as well as an annealing phase. In the warmup phase, the learning rate grows from the minimum to the maximum, while in the annealing phase, the learning rate decays from the maximum back to the minimum. Intuitively, at the beginning, we start with the minimum learning rate to ensure stability and avoid divergence, then in the middle, we want to move quickly toward a correct region of the parameters, with a large learning rate; at the end, we want to be careful again with a small learning rate because we want to fine-tune the parameters and search for an exact position that minimizes our loss.

1.4   PyTorch: An Example

Since the SGD algorithm heavily relies on computing gradients, it is crucial to use software that can perform differentiation effectively and efficiently. Fortunately, there are several open-source software options available. Among the most popular ones is PyTorch, an open-source package with a Python interface, mainly developed by Meta AI. It is a preferred choice for many researchers and practitioners in deep learning. To get started with PyTorch, you can visit the official PyTorch website at https://pytorch.org/. There, you will find installation instructions tailored to your specific operating system and tutorials that cover the basics of PyTorch, enabling you to quickly start building and training neural networks. Other popular choices for deep learning frameworks include TensorFlow, Keras, and JAX, each with its own unique features and benefits.

The experiments in this book are tested using PyTorch version 1.12.0. We can simply implement the following code to see the version:

```
import torch
print(f'torch version: {torch.__version__}')
```

We get the output with the version of PyTorch installed, similar to the following result:

```
torch version: 1.12.0
```

It is possible to define customized functions in PyTorch. For instance, the following defines the linear transformation and ReLU function for a PyTorch Tensor input x:

```
def lin(x, w, b):
    return x @ w + b

def relu(x):
    return x.clamp_min(0.)
```

Of course, we can imagine that PyTorch has many convenient built-in functions for constructing neural networks. The `torch.nn` class contains most of the building blocks. For instance, `torch.nn.Linear` and `torch.nn.ReLU` can easily implement the linear layer and ReLU activation.

Now, let's consider an example of image classification to showcase a typical training process using PyTorch. The dataset we use is the Fashion MNIST dataset available at https://github.com/zala ndoresearch/fashion-mnist. It contains 10 classes of fashion-related images. The 10 classes are T-shirt tops, trousers, pullovers, dresses, coats, sandals, shirts, sneakers, bags, and ankle boots. The classification task refers to assigning a class to each given input image. The dataset consists of a training set of 60,000 examples and a test set of 10,000 examples. All classes are approximately evenly distributed. Each image is in grayscale and has size 28×28. Figure 1.5 illustrates some examples of images from this dataset.

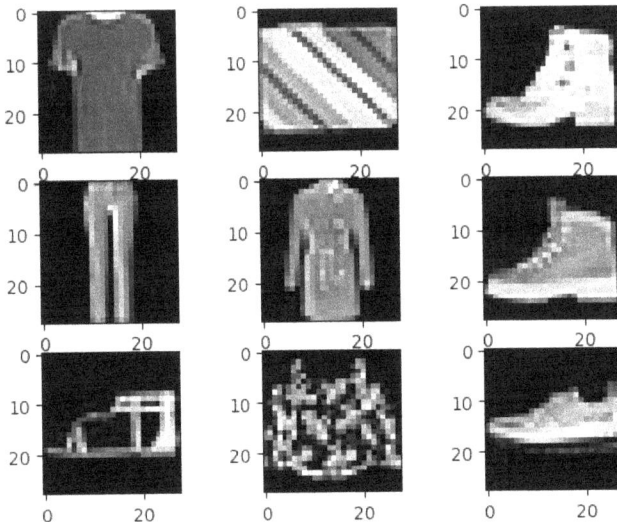

Fig. 1.5. Sample images from Fashion MNIST.

Instead of downloading the dataset directly from the website, we can also use the built-in PyTorch function to obtain the data. For instance, we can obtain the data by implementing the following Python codes:

```python
from torchvision import datasets
from torchvision.transforms import ToTensor

training_data = datasets.FashionMNIST(
    root="data", # we can customize the path
    train=True, # training data
    download=True, # once downloaded, we can set it to FALSE
    transform=ToTensor() # converting an image to a
        torch.FloatTensor with range [0., 1.]
)
test_data = datasets.FashionMNIST(
    root="data", # we can customize the path
    train=False, # test data
    download=True, # once downloaded, we can set it to FALSE
    transform=ToTensor() # converting an image to a
        torch.FloatTensor with range [0., 1.]
)
```

To tackle the classification problem, we construct a neural network that learns from the training data. In the case of a 10-class classification problem, our neural network will produce a 10-dimensional output vector, where each element represents the probability of the input data point belonging to a specific class (assuming we use numbers 1–10 to represent the class labels). More precisely, given an input $\mathbf{x}$ (e.g., a vectorized image in $\mathbb{R}^n$, where n is the total number of pixels in an image), the output is $\mathbf{z} = (z_1, \ldots, z_{10})^\top \in \mathbb{R}^{10}$. To interpret $\mathbf{z}$ as probabilities representing the likelihood of each class, we can transform it into $\tilde{\mathbf{z}} = (\tilde{z}_1, \ldots, \tilde{z}_{10})^\top \in \mathbb{R}^{10}$ such that

$$\tilde{z}_i \equiv \mathrm{softmax}_i(\mathbf{z}) := \frac{\exp(z_i)}{\sum_{j=1}^{10} \exp(z_j)}, \qquad (1.19)$$

to represent $\mathbb{P}(\text{class} = i | \mathbf{x})$. Here, the transformation $\tilde{\mathbf{z}} = \mathrm{softmax}(\mathbf{z})$ is widely known as the *softmax* function, which is commonly used in classification tasks to convert a vector into a probability distribution.

Note that the true label can also be represented as a 10-dimensional vector. Suppose the true class is k, then we can label it as a vector $\mathbf{y} \in \mathbb{R}^{10}$ such that $y_i = 1$ if $i = k$ and $y_i = 0$ if $i \neq k$.

There is only one element in $\mathbf{y}$ equal to 1 and all the other elements are equal to 0. Such a vector is called a *one-hot* vector, with the interpretation that 1 is hot and 0 is cold.

By comparing the output probability distribution $\tilde{\mathbf{z}}$ and the groundtruth one-hot vector $\mathbf{y}$, we can compute a cross-entropy (CE) loss for $\mathbf{x}$ as

$$\mathcal{E}_{\mathrm{CE}}(\mathbf{x}) = -\sum_{i=1}^{10} y_i \log(\tilde{z}_i). \tag{1.20}$$

Intuitively, the summation in (1.20) computes the similarity between $(y_1, \ldots, y_{10})$ and $(z_1, \ldots, z_{10})$ and therefore the negation can serve as a loss function. It is important to note that, since $\mathbf{y}$ is a one-hot vector, only one term in the summation is present. In PyTorch, we only need the neural network to output $\mathbf{z}$, called the *logits*.

You may wonder why the logit has such a name: It is because each logit will go through an exponential function when applying the softmax function. After the exponential function cancels "log", we are left with "it"! The loss function in PyTorch using `torch.nn.CrossEntropyLoss` will directly output the loss using the logits and the one-hot labels.

In PyTorch, we usually define neural networks as objects of classes. For instance, with the following codes, we can build a three-layer fully connected neural network:

```
from torch import nn

class NeuralNetwork(nn.Module):
    def __init__(self):
        super(NeuralNetwork, self).__init__()
        self.flatten = nn.Flatten() # vectorizing input images
        self.fcn = nn.Sequential(
            nn.Linear(28*28, 512), # first layer
            nn.ReLU(), # activation
            nn.Linear(512, 512), # second layer
            nn.ReLU(), # activation
            nn.Linear(512, 10), # third layer
            nn.ReLU() # activation
        )
    def forward(self, x):
        x = self.flatten(x)
        logits = self.fcn(x)
        return logits
```

Problem 1.1. How many parameters are in the above neural network?

In order to train a neural network in the above class efficiently, we may want to use a cuda GPU. The following line of codes sets the device to be GPU if a GPU is available:

```
device = 'cuda' if torch.cuda.is_available() else 'cpu'
```

We can then use the PyTorch dataloader to load the data and train the neural network. The following codes serve as an example:

```
from torch.utils.data import DataLoader
train_dataloader = DataLoader(training_data, batch_size=64) # a
    mini-batch contains 64 examples
test_dataloader = DataLoader(test_data, batch_size=64)

def train_loop(dataloader, model, loss_fn, optimizer):
    size = len(dataloader.dataset)
    for batch, (X, y) in enumerate(dataloader):
        # Compute prediction and loss
        X, y = X.to(device), y.to(device)
        pred = model(X) # logits, no need to apply softmax
        loss = loss_fn(pred, y)

        # Backpropagation (where training happens)
        optimizer.zero_grad()
        loss.backward()
        optimizer.step()

        '''
        # Use to monitor the training loss

        if batch % 100 == 0:
            loss, current = loss.item(), batch * len(X)
            print(f'loss: {loss:>7f} [{current:>5d}/{size:>5d}]')

        '''

def test_loop(dataloader, model, loss_fn):
    size = len(dataloader.dataset)
    num_batches = len(dataloader)
    test_loss, correct = 0, 0
```

```
   with torch.no_grad():
       for X, y in dataloader:
           X, y = X.to(device), y.to(device)
           pred = model(X)
           test_loss += loss_fn(pred, y).item()
           correct += (pred.argmax(1) ==
               y).type(torch.float).sum().item() # count the number
               of correct predictions: the prediction is correct
               each time the argmax of pred agrees with y

   test_loss /= num_batches
   correct /= size # accuracy rate = num of correct prediction
       divided by total num of instances
   print(f'Test Error: \n Accuracy: {(100*correct):>0.1f}%, Avg
       loss: {test_loss:>8f} \n')

model = NeuralNetwork().to(device)

loss_fn = nn.CrossEntropyLoss()

learning_rate = 1e-3
optimizer = torch.optim.SGD(model.parameters(), lr=learning_rate) #
    use SGD optimizer from torch.optim; need to specify the
    parameters and the learning rate

epochs = 5
for t in range(epochs):
    print(f'Epoch {t+1}\n-----------------------------')
    train_loop(train_dataloader, model, loss_fn, optimizer)
    test_loop(test_dataloader, model, loss_fn)
print("Complete!")
```

The following is an example of output of the above code. We see that the accuracy keeps improving with the epochs. At the end, the accuracy is over 50%, which is much better than a random guess (10% chance):

```
Epoch 1
-----------------------------
Test Error:
 Accuracy: 38.5%, Avg loss: 2.218973
```

```
Epoch 2
-------------------------------
Test Error:
 Accuracy: 44.0%, Avg loss: 2.092364
Epoch 3
-------------------------------
Test Error:
 Accuracy: 45.1%, Avg loss: 1.897768
Epoch 4
-------------------------------
Test Error:
 Accuracy: 47.0%, Avg loss: 1.703194
Epoch 5
-------------------------------
Test Error:
 Accuracy: 52.0%, Avg loss: 1.564697
Complete!
```

Problem 1.2. The 1-cycle scheduler can be called in PyTorch using `torch.optim.lr_scheduler.OneCycleLR`. For instance, we can define

```
scheduler = torch.optim.lr_scheduler.OneCycleLR(optimizer,
    max_lr=0.01, steps_per_epoch=len(train_dataloader), epochs=5)
```

When we use the scheduler, we need to add an additional `scheduler.step()` after `optimizer.step()` in the training loop. Try Modifying the code in the text to train a classifier for Fashion MNIST with the 1-cycle scheduler. You should observe a higher accuracy than the above result.

Problem 1.3. Train a neural network for the MNIST dataset available at `torchvision.datasets.MNIST`. Try other nonlinear activation functions introduced in this chapter.

Chapter 2

Neural Networks in Use

2.1 Differentiation

As discussed Chapter 1, modern deep learning frameworks heavily rely on auto-differentiation, which enables efficient computation of gradients. Initially, popular frameworks like Theano and Caffe emerged, providing tools for automatic differentiation and network training. However, as the field progressed, TensorFlow and PyTorch emerged as dominant players in the deep learning landscape (Caffe was merged into PyTorch). Additionally, if we only want to use high-level APIs, a good choice is Keras, which used to support several different backends but now completely relies on TensorFlow.

In the realm of neural networks, *backpropagation* is the algorithm used in the above frameworks for differentiating a neural network. It leverages the chain rule from calculus to compute derivatives in a structured and systematic manner. The architecture of a neural network plays a crucial role in enabling backpropagation, as it provides the necessary structure to track the dependencies among variables.

In calculus, the chain rule allows us to compute the derivative of a composite function by sequentially applying derivatives of its constituent functions. For instance, if we have $y = g(x)$ and $z = f(y)$, where x, y, and z are scalars, then

$$\frac{dz}{dx} = f'(g(x))g'(x) = \frac{dz}{dy}\frac{dy}{dx}. \tag{2.1}$$

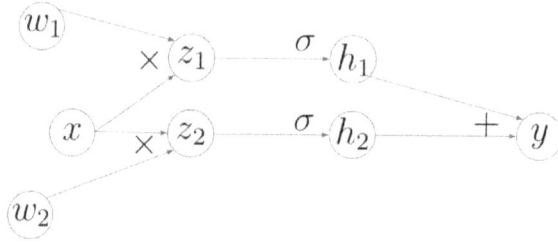

Fig. 2.1. A very simple neural network.

More generally, consider vectors $\mathbf{x}$, $\mathbf{y}$, and $\mathbf{z}$. Suppose $\mathbf{y} = g(\mathbf{x})$ and $\mathbf{z} = f(\mathbf{y})$. Then

$$\nabla_{\mathbf{x}} z = \nabla_{\mathbf{x}} (f \circ g)(\mathbf{x}) = [\nabla_{\mathbf{x}} g(\mathbf{x})]^{\top} \nabla_{\mathbf{y}} f(g(\mathbf{x})) = \mathbf{J}^{\top} \nabla_{\mathbf{y}} \mathbf{z}, \qquad (2.2)$$

where $\mathbf{J}$ is the Jacobian matrix whose (i, j)th entry is $J_{ij} = \partial y_i / \partial x_j$. It is important to note the transpose applied to the Jacobian. In the context of a neural network, we can interpret $\mathbf{x}$ as the input to the network, $\mathbf{z}$ as the output, and $\mathbf{y}$ as the hidden layer activations. When we desire to compute the derivative of $\mathbf{z}$ with respect to $\mathbf{x}$, we perform a backward traversal through the network, calculating intermediate derivatives along the way before taking their product.

To illustrate the idea, let's examine a very simple neural network, whose *computational graph* is illustrated in Figure 2.1. The computational graph showcases the sequence of mathematical operations that are applied to the input data to produce the desired output. In the given computational graph, we have an input node x, which is connected to two weight nodes w_1 and w_2. The weights w_1 and w_2 are multiplied with the input x to produce the intermediate nodes $z_1 = w_1 x$ and $z_2 = w_2 x$, respectively. These intermediate nodes z_1 and z_2 are then passed through an activation function σ, to produce the nodes h_1 and h_2. Finally, h_1 and h_2 are summed together to obtain the overall output y.

Suppose we want to compute $\partial y / \partial w_1$. We can perform a *forward pass* that goes from the input x to the output y so that we have all the numerical values at all the nodes. To compute the derivative, we can start with y and traverse back through the chain $y \rightarrow h_1 \rightarrow z_1 \rightarrow w_1$ and compute the derivatives along the way. As we move from y to h_1, we calculate $\partial y / \partial h_1 = 1$. Continuing from h_1 to z_1, we calculate $\partial h_1 / \partial z_1 = \sigma'(z_1)$. Finally, we move from z_1 to w_1 and calculate the corresponding derivative $\partial z_1/\partial w_1 = x$. To obtain the overall derivative $\partial y/\partial w_1$, we multiply all the derivatives along the path:

$$\frac{\partial y}{\partial w_1} = \frac{\partial y}{\partial h_1} \cdot \frac{\partial h_1}{\partial z_1} \cdot \frac{\partial z_1}{\partial w_1} = 1 \cdot \sigma'(z_1) \cdot x. \tag{2.3}$$

Similarly, if we want to compute $\partial y/\partial x$, we need to take into account the two paths $y \to h_1 \to z_1 \to x$ and $y \to h_2 \to z_2 \to x$, both from y to x. We thus need to take the sum of the derivatives along both paths. This process is convenient since the paths are determined by the architecture of the neural network. Starting from y, we traverse the first path $y \to h_1 \to z_1 \to x$. Along this path, we have $\partial y/\partial h_1 = 1$, $\partial h_1/\partial z_1 = \sigma'(z_1)$, and $\partial z_1/\partial x = w_1$. Therefore, the contribution to $\partial y/\partial x$ from this path is

$$\frac{\partial y}{\partial h_1} \cdot \frac{\partial h_1}{\partial z_1} \cdot \frac{\partial z_1}{\partial x} = 1 \cdot \sigma'(z_1) \cdot w_1.$$

Similarly, for the second path $y \to h_2 \to z_2 \to x$, we have $\partial y/\partial h_2 = 1$, $\partial h_2/\partial z_2 = \sigma'(z_2)$, and $\partial z_2/\partial x = w_2$. The contribution to $\partial y/\partial x$ from this path is

$$\frac{\partial y}{\partial h_2} \cdot \frac{\partial h_2}{\partial z_2} \cdot \frac{\partial z_2}{\partial x} = 1 \cdot \sigma'(z_2) \cdot w_2.$$

Finally, we sum up the contributions from both paths to obtain the total derivative $\partial y/\partial x = \sigma'(z_1)w_1 + \sigma'(z_2)w_2$. Figure 2.2 shows the calculation of derivatives during the above backpropagation.

There are two main approaches for performing backpropagation in auto-differentiation software: symbol-to-number differentiation and symbol-to-symbol differentiation. In symbol-to-number differentiation, the software takes the forward computational graph along with

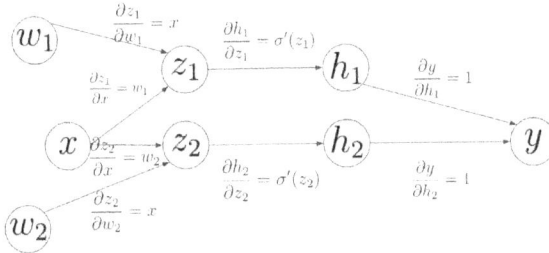

Fig. 2.2. Calculating the derivatives when performing backpropagation.

numerical values as inputs and computes a set of numerical values as outputs. This approach directly evaluates the derivatives using the given numerical values without creating a symbolic representation of the derivatives. On the other hand, in symbol-to-symbol differentiation, the software creates a symbolic description of the derivatives and constructs a backward computational graph for performing the backpropagation. For instance, originally, TensorFlow takes a symbol-to-symbol approach, where the computational graph was built before the forward pass, composed of symbols. In contrast, PyTorch takes a symbol-to-number approach, where the graph is created dynamically during the forward pass, with the numerical values.

Problem 2.1. Let's consider backpropagation for binary classification. Assume a neural network that outputs a logit $a_i \in \mathbb{R}$ for each input $\mathbf{x}_i$, $i = 1, \ldots, N$ of a batch. The overall cross-entropy loss is thus

$$\mathcal{E}_{\text{CE}} = -\sum_{i=1}^{N}(t_i \log y_i + (1 - t_i) \log(1 - y_i)), \tag{2.4}$$

where for each i, $t_i \in \{0, 1\}$ is the label corresponding to $\mathbf{x}_i$, and

$$y_i = \text{sigmoid}(a_i) = \frac{1}{1 + e^{-a_i}}. \tag{2.5}$$

Show that the derivative of the error function satisfies

$$\frac{\partial \mathcal{E}_{\text{CE}}}{\partial a_i} = y_i - t_i. \tag{2.6}$$

Compare the setting of binary classification with multi-class classification introduced in Chapter 1.

2.2 Regularization

In machine learning, regularization methods play a crucial role in incorporating prior beliefs into models. In the context of deep learning, traditional regularization techniques such as adding an explicit regularization term to penalize large parameter values can still be effective. However, there are also specialized regularization methods specifically designed for neural networks. In this section, we introduce several widely used techniques in deep learning regularization.

Batch normalization: Batch normalization refers to the operation of recentering and rescaling of outputs of neural network layers. It was originally proposed by Ioffe and Szegedy (2015) to improve the efficiency and stability of training neural networks. Its main motivation is to address the problem of internal covariate shift, which refers to the "change in the distribution of network activations due to the change in network parameters during training".

Problem 2.2. Train the same neural network defined in Chapter 1 for 50 iteration steps. Plot the histogram of activation values of the second linear layer versus iterations. This will give us an understanding of the distribution of activations without batch normalization. Next, add batch normalization (e.g., using the FCN defined in the following code box). Plot the histogram of activations after the second batch normalization versus iterations. This will allow us to observe the effect of batch normalization on the distribution of activations.

To address the issue of internal covariate shift, a common approach is to "standardize" or "whiten" the activations. This process is applied to individual neurons within a specific layer, treating each activation as a scalar. The same procedure is performed simultaneously for all neurons. During training, the whitening operation is done over a mini-batch $\mathbb{B} = \{x_1, \ldots, x_m\}$ of activations. That is,

$$\hat{x}_i = \frac{x_i - \mu_{\mathbb{B}}}{\sqrt{\sigma_{\mathbb{B}}^2 + \epsilon}}, \tag{2.7}$$

where $\mu_{\mathbb{B}} = \frac{1}{m}\sum_{i=1}^{m} x_i$ and $\sigma_{\mathbb{B}}^2 = \frac{1}{m}\sum_{i=1}^{m}(x_i - \mu_{\mathbb{B}})^2$ are the minibatch mean and the mini-batch variance, respectively. However, the procedure does not end with whitening the activations. In batch normalization, we still need to learn two parameters: the new scale γ and the new center β (the parameters are different for each neuron in the layer). These parameters allow us to control the scaling and shifting of the normalized activations. The output of batch normalization can be described as follows:

$$y_i = \gamma\hat{x}_i + \beta. \tag{2.8}$$

It is custom to consider batch normalization as a layer in a neural network. For instance, the following code inserts batch normalization layers into the fully connected network that we have seen in Chapter 1.

```
from torch import nn

self.fcn_with_bn = nn.Sequential(
    nn.Linear(28*28, 512),
    nn.ReLU(),
    nn.BatchNorm1d(512), # batch normalization layer
    nn.Linear(512, 512),
    nn.ReLU(),
    nn.BatchNorm1d(512), # batch normalization layer
    nn.Linear(512, 10),
    nn.ReLU()
)
```

In the above, the `BatchNorm1d` function is used because the input consists of one-dimensional vectors. However, in scenarios where the input is two-dimensional, such as images, we will need to use the `BatchNorm2d` function.

Dropout: Dropout randomly sets some activations to zero with a specified probability p during training. If we apply dropout with probability p to a layer, then each neuron of the layer is present with a probability of $1 - p$ at training time. At test time, all the neurons are present but the weights are scaled by $1 - p$. Equivalently, in PyTorch for instance, the activations are scaled by $(1-p)^{-1}$ during training, and no additional adjustments are needed at test time.

The use of dropout can be seen as a form of model averaging, where the network learns to make predictions by considering multiple possible subnetworks (realizations of the random dropout). This helps improve robustness of the model against noise and enhances its ability to generalize well to unseen data.

Similar to batch normalization, we can also treat dropout as a layer of a neural network. For instance, in PyTorch, dropout layers can be stacked with other layers in the following way by modifying the codes for batch normalization. Note that we need to specify the probability p in the code. If we leave it as blank, then $p = 0.5$ will be used by default.

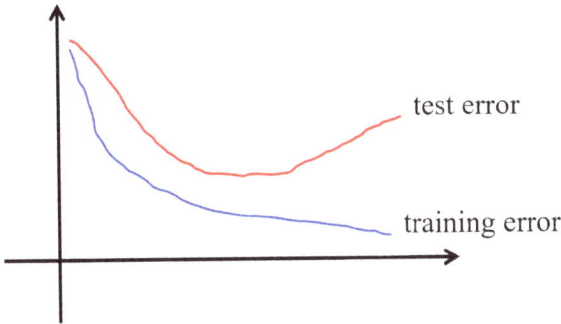

Fig. 2.3. Over-training causes generalization error.

```
from torch import nn

self.fcn_with_dropout = nn.Sequential(
    nn.Linear(28*28, 512),
    nn.ReLU(),
    nn.Dropout(p=0.2), # dropout layer with p=0.2
    nn.Linear(512, 512),
    nn.ReLU(),
    nn.Dropout(), # dropout layer with p=0.5
    nn.Linear(512, 10),
    nn.ReLU()
)
```

Early stopping: Early stopping is a simple and effective way to prevent overfitting and improve generalization. As shown in Figure 2.3, if we over train a neural network, we may get a very small training error, but it generalizes terribly with a large test error. In this case, during the training process, we should monitor the validation error, the loss value from the validation data. At the beginning of the training process, the validation error should decrease together with the training error. At a certain step, it may start increasing. By observing this increase in validation error, we can stop the training process early and use the model with the best validation performance as our final model.

Of course, regularization techniques commonly employed in machine learning can also be applied to training neural networks.

An example is L^2 regularization, which we review in the following exercise.

Problem 2.3. Suppose the objective function in training is $J(\boldsymbol{\theta}; \mathbf{X}, \mathbf{y})$, where $\boldsymbol{\theta}$ represents the parameters, $\mathbf{X}$ is the training data input, and $\mathbf{y}$ is target data output. One commonly used regularization is L^2 regularization, where the objective function is

$$\tilde{J}(\boldsymbol{\theta}; \mathbf{X}, \mathbf{y}) = J(\boldsymbol{\theta}; \mathbf{X}, \mathbf{y}) + \frac{\alpha}{2} \boldsymbol{\theta}^\top \boldsymbol{\theta}, \tag{2.9}$$

whose gradient with respect to $\boldsymbol{\theta}$ is

$$\nabla_{\boldsymbol{\theta}} \tilde{J}(\boldsymbol{\theta}; \mathbf{X}, \mathbf{y}) = \nabla_{\boldsymbol{\theta}} J(\boldsymbol{\theta}; \mathbf{X}, \mathbf{y}) + \alpha \boldsymbol{\theta}. \tag{2.10}$$

(1) Write the iteration step in the gradient descent method.
(2) Assume the objective function is quadratic and $\boldsymbol{\theta}^* = \arg\min_{\boldsymbol{\theta}} J(\boldsymbol{\theta})$, then

$$J(\boldsymbol{\theta}) = J(\boldsymbol{\theta}^*) + \frac{1}{2} (\boldsymbol{\theta} - \boldsymbol{\theta}^*)^\top \mathbf{H} (\boldsymbol{\theta} - \boldsymbol{\theta}^*).$$

Show that the minimizer of $\tilde{J}$ is given by

$$\tilde{\boldsymbol{\theta}} = (\mathbf{H} + \alpha \mathbf{I})^{-1} \mathbf{H} \boldsymbol{\theta}^*.$$

(3) Show that $\tilde{\boldsymbol{\theta}}$ is a scaled version of $\boldsymbol{\theta}^*$ in the sense that along the direction of the ith eigenvector of $\mathbf{H}$, $\boldsymbol{\theta}^*$ is scaled by a factor of

$$\frac{\lambda_i}{\lambda_i + \alpha}.$$

2.3   Universal Approximation

One of the key reasons for the widespread application of neural networks is their expressivity, which refers to their ability to approximate any continuous function. Neural networks form a powerful hypothesis class, capable of representing a wide range of complex functions. In this section, we try to understand the reason behind expressivity of neural networks.

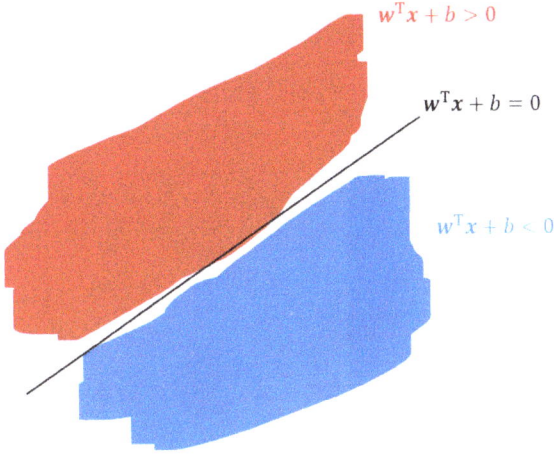

$\mathbf{w}^\mathsf{T}\mathbf{x} + b > 0$

$\mathbf{w}^\mathsf{T}\mathbf{x} + b = 0$

$\mathbf{w}^\mathsf{T}\mathbf{x} + b < 0$

Fig. 2.4. Illustration of $\text{Sign}(\mathbf{w}^\mathsf{T}\mathbf{x} + b)$. The hyperplane $\{\mathbf{x} \in \mathbb{R}^n : \mathbf{w}^\mathsf{T}\mathbf{x} + b = 0\}$ splits the space $\mathbb{R}^n$ into two parts. The red part corresponds to $\mathbf{w}^\mathsf{T}\mathbf{x} + b > 0$ and thus $\text{Sign}(\mathbf{w}^\mathsf{T}\mathbf{x} + b) = 1$, whereas the blue part corresponds to $\mathbf{w}^\mathsf{T}\mathbf{x} + b < 0$ and thus $\text{Sign}(\mathbf{w}^\mathsf{T}\mathbf{x} + b) = -1$.

Let's first look at the simplest activation function

$$\sigma(x) = \text{Sign}(x) = \begin{cases} 1, & \text{if } x > 0, \\ -1, & \text{otherwise} \end{cases} \tag{2.11}$$

for the moment. If we apply σ to some $\mathbf{w}^\mathsf{T}\mathbf{x} + b$, the resulting function $\sigma(\mathbf{w}^\mathsf{T}\mathbf{x} + b)$ will have as its range binary values $\{\pm 1\}$. Figure 2.4 clearly shows that the space of input is partitioned into two parts according to the outputs.

The simplest conclusion we can draw, with this sign activation function, is a theorem that describes expressivity of fully connected networks with discrete input vectors.

Theorem 2.1. *Let $n \in \mathbb{N}$. There exists a collection of fully connected networks that contains all functions from $\{\pm 1\}^n$ to $\{\pm 1\}$.*

Proof. Fix a function $f : \{\pm 1\}^n \to \{\pm 1\}$. We construct a neural network to represent f. Let $\mathbb{S} = \{\mathbf{u}_1, \ldots, \mathbf{u}_K\}$ denote the set of all input vectors with which the output of f is 1. For any $j \in \{1, \ldots, K\}$, and any $\mathbf{x} \in \{\pm 1\}^n$, we have the following:

- $\mathbf{u}_j^\mathsf{T} \mathbf{x} = n$, if $\mathbf{x} = \mathbf{u}_j$;
- $\mathbf{u}_j^\mathsf{T} \mathbf{x} \leq n - 2$, if $\mathbf{x} \neq \mathbf{u}_j$.

Let's define $g_j(\mathbf{x}) = \text{Sign}(\mathbf{u}_j^\top \mathbf{x} - (n-1))$. Obviously,

$$g_j(\mathbf{x}) = \begin{cases} 1, & \text{if } \mathbf{x} = \mathbf{u}_j, \\ -1, & \text{if } \mathbf{x} \neq \mathbf{u}_j. \end{cases}$$

In other words, g_j acts as an indicator of the member $\mathbf{u}_j$ in $\mathbb{S}$. To represent f, we only need to represent the set $\mathbb{S}$. To achieve this, we take the sum of all g_j's:

$$h(\mathbf{x}) = \sum_{j=1}^{K} g_j(\mathbf{x}).$$

Then,

$$h(\mathbf{x}) = \begin{cases} 1 \cdot 1 + (-1)(n-1) = -n + 2, & \text{if } \mathbf{x} \in \mathbb{S}, \\ (-1)(n) = -n, & \text{elsewhere}. \end{cases}$$

Consider $\text{NN}(\mathbf{x}) = h(\mathbf{x}) + (n-1)$. It equals 1 if $\mathbf{x} \in \mathbb{S}$ and -1 if $\mathbf{x} \notin \mathbb{S}$. That exactly means $\text{NN} = f$. On the other hand, NN is clearly a neural network whose first layer is given by g_j's and second layer by a simple summation with bias $(n-1)$. $\qquad\square$

Now let's keep using the sign activation function but with a continuous domain, say $\mathbb{R}^n$. The value of the sign function depends on the side of the hyperplane $\mathbf{w}^\top \mathbf{x} + b = 0$, where $\mathbf{x}$ is located.

Of course, a single layer cannot be very expressive. Suppose we want to express something more complicated, for instance, the indicator function of a triangle in $\mathbb{R}^2$. That is, we want to output 1 if the input is inside the triangle and output 0 otherwise. In this case, we need more than one layer. First, we consider the hyperplanes (one-dimensional lines) that contain the three sides of the triangle. Say these hyperplanes are represented by the equations $\mathbf{w}_1^\top \mathbf{x} + b_1 = 0$, $\mathbf{w}_2^\top \mathbf{x} + b_2 = 0$, and $\mathbf{w}_3^\top \mathbf{x} + b_3 = 0$, respectively, where $\mathbf{w}_j \in \mathbb{R}^2$, $j = 1, 2, 3$, such that $g_j(\mathbf{x}) = \sigma(\mathbf{w}_j \mathbf{x} + b_j)$ takes the value 1 on the side that contains the triangle and takes the value -1 on the opposite side, for $j = 1, 2, 3$. Next, let's consider the sum $g_1 + g_2 + g_3$, whose values are illustrated in Figure 2.5.

Clearly, inside the triangle, we have $g_1 + g_2 + g_3 = 3$. On the other hand, outside the triangle, we have either $g_1 + g_2 + g_3 = 1$ or

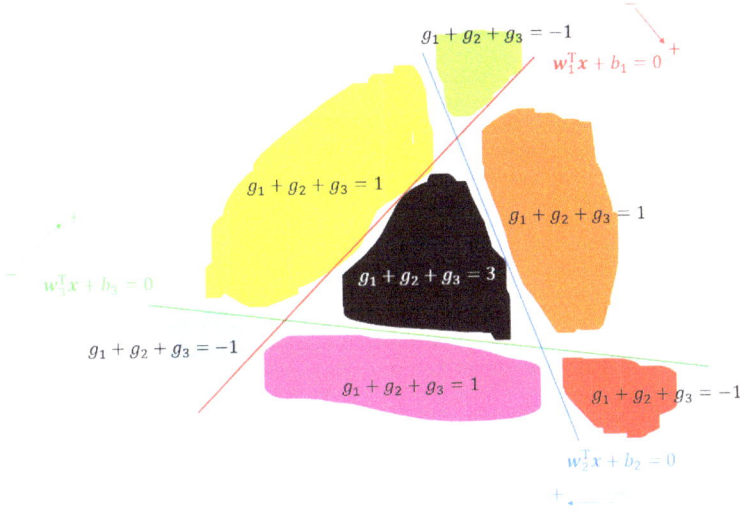

Fig. 2.5. Illustration of $g_1 + g_2 + g_3$. The signs of $\mathbf{w}_i^\top \mathbf{x} + b_i$, $i = 1, 2, 3$ are indicated by the respective arrows. The region inside the triangle is the only region where $g_1 + g_2 + g_3 > 2$.

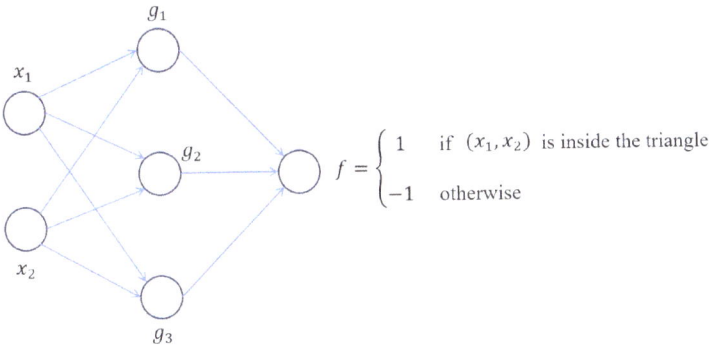

$$f = \begin{cases} 1 & \text{if } (x_1, x_2) \text{ is inside the triangle} \\ -1 & \text{otherwise} \end{cases}$$

Fig. 2.6. The indicator function of the triangle.

$g_1 + g_2 + g_3 = 0$. Let's now look at the function

$$f(\mathbf{x}) = \sigma(g_1(\mathbf{x}) + g_2(\mathbf{x}) + g_3(\mathbf{x}) - 2). \tag{2.12}$$

Then $f(\mathbf{x}) = 1$ if $\mathbf{x}$ is inside the triangle and $f(\mathbf{x}) = 0$ if $\mathbf{x}$ is outside. We can consider f as a two-layer neural network as illustrated in Figure 2.6.

With a similar approach, we can build neural networks that represent indicator functions for any polygons. These indicator functions can be combined to construct more complicated functions. Therefore, it is no surprise that neural networks have very strong expressive power!

Problem 2.4. According to the above discussion, design a neural network that implements the "XOR" function (cf. Figure 1.1).

For more complex functions, the expressive power of neural networks is often summarized by the universal approximation theorem, which states that any continuous function from $[-1,1]^n$ to $\mathbb{R}$ can be well approximated by a fully connected neural network. The original result requires some advanced mathematical tool, but it will be much easier if we are not too ambitious and only want to approximate Lipschitz continuous functions. We summarize the property in the following exercise.

Problem 2.5. Let $\mathbb{K}$ be a closed and bounded domain in $\mathbb{R}^n$. Let $f : \mathbb{K} \to \mathbb{R}$ be a Lipschitz continuous function such that

$$|f(\mathbf{x}) - f(\mathbf{y})| \leq L \, \|\mathbf{x} - \mathbf{y}\| \text{ for any } \mathbf{x}, \mathbf{y} \in \mathbb{K}. \qquad (2.13)$$

Then, given any $\epsilon > 0$, there exists a neural network NN : $\mathbb{K} \to \mathbb{R}$ such that

$$|f(\mathbf{x}) - \mathrm{NN}(\mathbf{x})| \leq \epsilon, \qquad (2.14)$$

for any $\mathbf{x} \in \mathbb{K}$.

The idea of proving Problem 2.5 is as follows. First, we partition $\mathbb{K}$ into very small regions. For $\mathbf{x}$ and $\mathbf{y}$ in the same region, $\|\mathbf{x} - \mathbf{y}\|$ will be small and so is $|f(\mathbf{x}) - f(\mathbf{y})|$ due to the Lipschitz condition (2.13). That is, on each region, f is almost constant. Then, for each small region, we can build a neural network that is the indicator function of the region multiplied by the constant function value corresponding to the region. Finally, we sum up the neural networks for all the regions.

Chapter 3

Optimization Methods

3.1 Gradient Descent

In Chapter 1, we briefly mentioned using gradient descent and stochastic gradient descent (SGD) for training neural networks. In this section, we first try to understand why moving the parameters along the direction of the negative gradient will decrease the function value.

Consider a typical optimization task we face in deep learning, say

$$\min_{\mathbf{w} \in \mathbb{R}^d} f(\mathbf{w}), \tag{3.1}$$

where f is the objective function depending on training data and $\mathbf{w}$ stands for the parameters. Following the gradient descent method, suppose we take T gradient steps in total. The tth gradient step can be written as

$$\mathbf{w}^{(t+1)} = \mathbf{w}^{(t)} - \eta^{(t)} \nabla f(\mathbf{w}^{(t)}), \ t = 0, 1, \dots, T - 1, \tag{3.2}$$

where we allow the learning rate $\eta^{(t)}$ to change over time.

An intuition why (3.2) works is the following. The first-order Taylor approximation of $f(\mathbf{w}^{(t+1)})$ is given by

$$f(\mathbf{w}^{(t+1)}) = f(\mathbf{w}^{(t)} - \eta^{(t)} \nabla f(\mathbf{w}^{(t)})) \tag{3.3}$$

$$\approx f(\mathbf{w}^{(t)}) - \eta^{(t)} \nabla f(\mathbf{w}^{(t)})^\top \nabla f(\mathbf{w}^{(t)}) \tag{3.4}$$

$$= f(\mathbf{w}^{(t)}) - \eta^{(t)} \|\nabla f(\mathbf{w}^{(t)})\|^2. \tag{3.5}$$

Therefore, as long as $\eta^{(t)}$ is small (so that the Taylor approximation does not produce a big error) and positive, we always have $f(\mathbf{w}^{(t+1)}) \leq f(\mathbf{w}^{(t)})$.

To strengthen our argument, let's provide further evidence supporting our claims. For the sake of simplicity, let's assume that the eigenvalues of the Hessian matrix $\nabla^2 f(\mathbf{w})$ have a uniform upper bound β. That is, any eigenvalue λ of $\nabla^2 f(\mathbf{w})$, regardless of the value of $\mathbf{w}$, satisfies $\lambda \leq \beta$. According to Taylor's theorem,

$$f(\mathbf{w}^{(t+1)}) = f(\mathbf{w}^{(t)} - \eta^{(t)}\nabla f(\mathbf{w}^{(t)})) \tag{3.6}$$

$$= f(\mathbf{w}^{(t)}) - \eta^{(t)}\|\nabla f(\mathbf{w}^{(t)})\|^2$$

$$+ \frac{(\eta^{(t)})^2}{2}\nabla f(\mathbf{w}^{(t)})^\top \nabla^2 f(\boldsymbol{\xi})\nabla f(\mathbf{w}^{(t)}), \tag{3.7}$$

where $\boldsymbol{\xi}$ is a vector lying on the line segment connecting $\mathbf{w}^{(t)}$ and $\mathbf{w}^{(t+1)}$. According to our assumption that the eigenvalues of $\nabla^2 f(\boldsymbol{\xi})$ are bounded above by β, we have

$$\nabla f(\mathbf{w}^{(t)})^\top \nabla^2 f(\boldsymbol{\xi})\nabla f(\mathbf{w}^{(t)}) \leq \beta\|\nabla f(\mathbf{w}^{(t)})\|^2. \tag{3.8}$$

Hence, plugging (3.8) into (3.7), we have

$$f(\mathbf{w}^{(t+1)}) \leq f(\mathbf{w}^{(t)}) - \eta^{(t)}\|\nabla f(\mathbf{w}^{(t)})\|^2 + \frac{\beta(\eta^{(t)})^2}{2}\|\nabla f(\mathbf{w}^{(t)})\|^2. \tag{3.9}$$

Taking $\eta^{(t)} = 1/\beta$ yields

$$f(\mathbf{w}^{(t+1)}) \leq f(\mathbf{w}^{(t)}) - \frac{1}{2\beta}\|\nabla f(\mathbf{w}^{(t)})\|^2. \tag{3.10}$$

Therefore, the value of f indeed decreases with the gradient steps.

3.1.1   *Analysis of convergence for gradient descent*

Case 1: Strong convexity.   We first assume that the eigenvalues of the Hessian matrix $\nabla^2 f(\mathbf{w})$ have a uniform lower bound $\alpha > 0$. This is indeed a very strong assumption and we say that

f is strongly convex because of the lower bound α. Furthermore, let $\mathbf{w}^* \in \arg\min f(\mathbf{w})$ be a minimizer of f. We have

$$f(\mathbf{w}^*) = f(\mathbf{w}^{(t)} + \mathbf{w}^* - \mathbf{w}^{(t)}) \tag{3.11}$$

$$= f(\mathbf{w}^{(t)}) + \nabla f(\mathbf{w}^{(t)})^\top (\mathbf{w}^* - \mathbf{w}^{(t)})$$

$$+ \frac{1}{2}(\mathbf{w}^* - \mathbf{w}^{(t)})^\top \nabla^2 f(\boldsymbol{\xi}^*)(\mathbf{w}^* - \mathbf{w}^{(t)}) \tag{3.12}$$

$$\geq f(\mathbf{w}^{(t)}) + \nabla f(\mathbf{w}^{(t)})^\top (\mathbf{w}^* - \mathbf{w}^{(t)}) + \frac{1}{2}\alpha \|\mathbf{w}^* - \mathbf{w}^{(t)}\|^2, \tag{3.13}$$

where $\boldsymbol{\xi}^*$ lies on the line segment connecting $\mathbf{w}^{(t)}$ and $\mathbf{w}^*$, and the last inequality follows from the fact that α is a lower bound of the eigenvalues of $\nabla^2 f(\boldsymbol{\xi})$. Next, we consider the function

$$g(\mathbf{w}^*) = f(\mathbf{w}^{(t)}) + \nabla f(\mathbf{w}^{(t)})^\top (\mathbf{w}^* - \mathbf{w}^{(t)}) + \frac{1}{2}\alpha \|\mathbf{w}^* - \mathbf{w}^{(t)}\|^2, \tag{3.14}$$

which is quadratic in $\mathbf{w}^*$ and we can easily find the optimal $\mathbf{w}^*$ by setting $\nabla_{\mathbf{w}^*} g(\mathbf{w}^*)$ to zero. That will give $\mathbf{w}^* = \mathbf{w}^{(t)} - \frac{\nabla f(\mathbf{w}^{(t)})}{\alpha}$ and the corresponding $g(\mathbf{w}^*) = f(\mathbf{w}^{(t)}) - \frac{\|\nabla f(\mathbf{w}^{(t)})\|^2}{2\alpha}$. To summarize, this implies

$$f(\mathbf{w}^*) \geq f(\mathbf{w}^{(t)}) - \frac{1}{2\alpha}\|\nabla f(\mathbf{w}^{(t)})\|^2. \tag{3.15}$$

Canceling the $\|\nabla f(\mathbf{w}^{(t)})\|^2$ terms in (3.10) and (3.15), we have

$$\beta f(\mathbf{w}^{(t+1)}) - \alpha f(\mathbf{w}^*) \leq (\beta - \alpha)f(\mathbf{w}^{(t)}), \tag{3.16}$$

or equivalently,

$$f(\mathbf{w}^{(t+1)}) - f(\mathbf{w}^*) \leq \left(1 - \frac{\alpha}{\beta}\right)(f(\mathbf{w}^{(t)}) - f(\mathbf{w}^*)). \tag{3.17}$$

This iterative relation implies

$$f(\mathbf{w}^{(T)}) - f(\mathbf{w}^*) \leq \left(1 - \frac{\alpha}{\beta}\right)^T (f(\mathbf{w}^{(0)}) - f(\mathbf{w}^*)). \tag{3.18}$$

We see that with the strong convexity assumption, the gradient descent method has a rapid convergence.

Case 2: Convexity. Now, we only assume that f is convex so that $f(\mathbf{w}) \leq f(\mathbf{w}') + \nabla f(\mathbf{w})^\top (\mathbf{w} - \mathbf{w}')$ for any $\mathbf{w}$ and $\mathbf{w}'$. In this case, we have

$$f(\mathbf{w}^{(t)}) \leq f(\mathbf{w}^*) + \nabla f(\mathbf{w}^{(t)})^\top (\mathbf{w}^{(t)} - \mathbf{w}^*). \tag{3.19}$$

Combining (3.19) and (3.10), we have

$$f(\mathbf{w}^{(t+1)})$$

$$\leq f(\mathbf{w}^*) + \nabla f(\mathbf{w}^{(t)})^\top (\mathbf{w}^{(t)} - \mathbf{w}^*) - \frac{1}{2\beta} \|\nabla f(\mathbf{w}^{(t)})\|^2 \tag{3.20}$$

$$= f(\mathbf{w}^*) + \frac{\beta}{2} \left(\|\mathbf{w}^{(t)} - \mathbf{w}^*\|^2 - \left\| \mathbf{w}^{(t)} - \mathbf{w}^* - \frac{1}{\beta} \nabla f(\mathbf{w}^{(t)}) \right\|^2 \right) \tag{3.21}$$

$$= f(\mathbf{w}^*) + \frac{\beta}{2} (\|\mathbf{w}^{(t)} - \mathbf{w}^*\|^2 - \|\mathbf{w}^{(t+1)} - \mathbf{w}^*\|^2). \tag{3.22}$$

Therefore, since $f(\mathbf{w}^{(t)})$ decreases with t, we have

$$f(\mathbf{w}^{(T)}) - f(\mathbf{w}^*)$$

$$\leq \frac{1}{T} \sum_{t=0}^{T-1} (f(\mathbf{w}^{(t+1)}) - f(\mathbf{w}^*)) \tag{3.23}$$

$$\leq \frac{1}{T} \sum_{t=0}^{T-1} \frac{\beta}{2} (\|\mathbf{w}^{(t)} - \mathbf{w}^*\|^2 - \|\mathbf{w}^{(t+1)} - \mathbf{w}^*\|^2) \tag{3.24}$$

$$= \frac{\beta}{2T} \left(\sum_{t=0}^{T-1} \|\mathbf{w}^{(t)} - \mathbf{w}^*\|^2 - \sum_{t=1}^{T} \|\mathbf{w}^{(t)} - \mathbf{w}^*\|^2 \right) \tag{3.25}$$

$$\leq \frac{\beta}{2T} \|\mathbf{w}^{(0)} - \mathbf{w}^*\|^2. \tag{3.26}$$

The last term is a constant times $1/T$. To emphasize the speed of convergence, we can also say that $f(\mathbf{w}^{(T)}) - f(\mathbf{w}^*) = O(1/T)$.

Similar analysis can also be performed for SGD, with slightly different assumptions. It is important to note that SGD and the batch setting do not necessarily have the same convergence rate. For instance, with an appropriate learning rate, the convergence rate for SGD is $O(1/\sqrt{T})$ (Shalev-Shwartz and Ben-David, 2014, Chapter 14).

The stochastic gradient descent can be implemented in PyTorch using the following code. As we have seen in Chapter 1, an optimizer is defined as an object which controls the change of the parameters:

```
import torch

# an optimizer is defined for a neural network, coded as an object
    called "model", the learning rate "lr" has to be specified
optimizer = torch.optim.SGD(model.parameters(), lr=1e-3)
# if we want to perform maximization, i.e., gradient ascent, we can
    simply specify "maximize=True"


    ''' 
    # the following three lines are implemented in training
        iterations
    optimizer.zero_grad()
    loss.backward()
    optimizer.step()
    ''' 
```

3.2   Momentum

One problem with gradient descent is that it may show zigzag tracks and does not converge quickly enough. Figure 3.1 illustrates a typical case even if the objective function is convex.

To overcome the inefficiency from the zigzag updates, a *momentum* method remembers the gradient direction in the previous step. That is, there is an inertia for the travel of $\mathbf{w}$ during iterations. For instance, the simplest momentum method applies the following update:

$$\phi^{(t+1)} = \phi^{(t)} + \gamma^{(t)} \nabla f(\mathbf{w}^{(t)}), \qquad (3.27)$$

$$\mathbf{w}^{(t+1)} = \mathbf{w}^{(t)} - \eta^{(t)} \phi^{(t+1)}. \qquad (3.28)$$

In (3.27), $\phi^{(t)}$ is the direction of update in the tth step (considered as driven by the inertia), whereas $\nabla f(\mathbf{w}^{(t)})$ is the gradient (considered as driven by the new force). Therefore, (3.27) calculates a linear

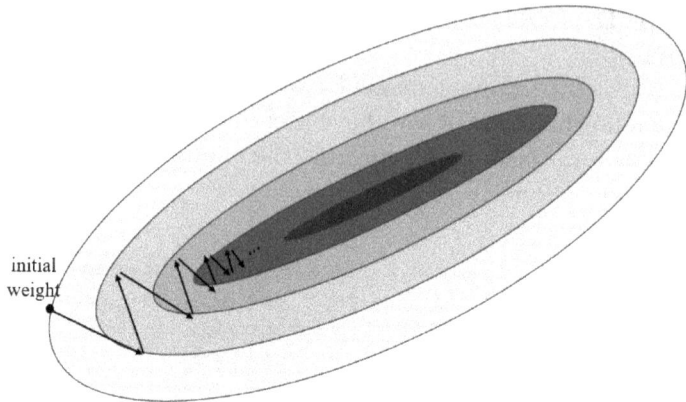

Fig. 3.1. The zigzag updates of gradient descent. The gradients of the objective function vary significantly across the two dimensions parallel to the axes of symmetry of the ellipse. As a result, gradient descent struggles to find a consistent direction of descent and instead oscillate between different directions.

combination of these two directions, which is then used as the update direction in (3.28).

A special and famous example of momentum method is known as *Nesterov's acceleration*, of which the update step is

$$\phi^{(t+1)} = \mathbf{w}^{(t)} - \gamma^{(t)} \nabla f(\mathbf{w}^{(t)}), \tag{3.29}$$

$$\mathbf{w}^{(t+1)} = \phi^{(t+1)} + \frac{t}{t+3}(\phi^{(t+1)} - \phi^{(t)}). \tag{3.30}$$

The term $t/(t+3)$ in (3.30) is not very intuitive and instead is quite technical. Let's omit the technical proof of Nesterov's acceleration, but note the fact that unlike the gradient descent method $f(\mathbf{w}^{(T)}) - f(\mathbf{w}^*) = O(1/T)$, Nesterov's acceleration enjoys faster convergence $f(\mathbf{w}^{(T)}) - f(\mathbf{w}^*) = O(1/T^2)$. Thus, if we don't have very good conditions such as strong convexity with which gradient descent converges exponentially fast, Nesterov's acceleration is a faster algorithm than gradient descent.

Despite the less intuitive update equation for Nesterov's acceleration, Su *et al.* (2014) provided an interpretation of Nesterov's acceleration from the perspective of ordinary differential equations (ODEs). Indeed, we can consider $\{\mathbf{w}^{(t)}\}_{t=0}^{T}$ as a discretization of $\mathbf{w}(s)$, where s can be seen as a continuous time variable. Here, $\mathbf{w}^{(t)} = \mathbf{w}(s^{(t)})$, and

$\eta^{(t)} = s^{(t+1)} - s^{(t)}$ is the time step between the time stamps $s^{(t+1)}$ and $s^{(t)}$. Then, (3.2) is consistent with the following ODE:

$$\frac{d\mathbf{w}(s)}{ds} = -\nabla f(\mathbf{w}(s)) \tag{3.31}$$

as $\eta^{(t)} \to 0$. Similarly, Nesterov's acceleration is consistent with the following ODE:

$$\frac{d^2\mathbf{w}(s)}{ds^2} + \frac{3}{s}\frac{d\mathbf{w}(s)}{ds} = -\nabla f(\mathbf{w}(s)). \tag{3.32}$$

This ODE serves as a better analytical tool for understanding Nesterov's acceleration, where $3/s$ serves as a damping ratio. If we change 3 to a different number r, it will behave differently when $r < 3$ and $r > 3$. When $r < 3$, the damping ratio is smaller, which means there is less damping or resistance to the oscillations in the algorithm. As a result, the algorithm may exhibit more oscillatory behavior, causing slower convergence or even instability. On the other hand, when $r > 3$, the damping ratio is larger, indicating a higher degree of damping. This increased damping helps suppress the oscillations, leading to faster convergence and more stable behavior of Nesterov's acceleration.

Problem 3.1. Validate (3.31).

In PyTorch, it is possible to implement a stochastic gradient method with momentum or Nesterov's acceleration. For instance, it is possible to implement the following code in replacement of a standard stochastic gradient:

```
optimizer = torch.optim.SGD(model.parameters(), lr=1e-3,
    momentum=0.8, nesterov=True)
```

In the above code, the "momentum" parameter corresponds to $1/\gamma^{(t)}$ in (3.27) and needs to be constant over the time steps. It is also worth noting that PyTorch implements Nesterov's acceleration slightly differently than the original formulation, using the same "momentum" parameter. In practice, to choose a best configuration, there is no "one-size-fits-all" answer, as the effectiveness of each technique can vary depending on the dataset, model architecture, and optimization landscape. We should compare their performance on the validation set to make an informed decision. In the context of deep learning, there are other widely used optimization algorithms, which we continue to introduce in the following section.

3.3 AdaGrad, RMSProp, and Adam

Another technique for accelerating convergence of gradient descent methods is preconditioning, where a $d \times d$ matrix $\mathbf{H}^{(t)}$ is used to modify the gradient step as

$$\mathbf{w}^{(t+1)} = \mathbf{w}^{(t)} - \eta^{(t)}(\mathbf{H}^{(t)})^{-1}\nabla f(\mathbf{w}^{(t)}). \tag{3.33}$$

To see why this may accelerate convergence, let's consider the example where the objective function is defined as $f(\mathbf{w}) = \frac{1}{2}\mathbf{w}^\top \mathbf{A}\mathbf{w} + \mathbf{b}^\top \mathbf{w} + c$. Let's try to be ambitious and require a one-step convergence: Let $\mathbf{w}^{(t)} \in \mathbb{R}^d$ be an arbitrary point. Let $\eta^{(t)} = 1$. We leave this as the following exercise.

Problem 3.2. Prove that $\nabla f(\mathbf{w}^{(t+1)}) = 0$ if and only if $\mathbf{H}^{(t)} = \mathbf{A}$ in (3.33).

Of course, in practice, we don't deal with quadratic functions. A more practical scheme of preconditioner is called *AdaGrad*, or adaptive preconditioner. Specifically,

$$\mathbf{G}^{(t)} = \left(\sum_{i=0}^{t} \nabla f(\mathbf{w}^{(i)})\nabla f(\mathbf{w}^{(i)})^\top\right)^{\frac{1}{2}}, \tag{3.34}$$

$$\mathbf{w}^{(t+1)} = \mathbf{w}^{(t)} - \eta(\mathbf{G}^{(t)})^{-1}\nabla f(\mathbf{w}^{(t)}). \tag{3.35}$$

Note that it is numerically impractical to compute (3.34) and (3.35), particularly due to the operations of taking the square root and the inverse of matrices. A more practical implementation, which is also implemented in `torch.optim.Adagrad`, is to take $\mathbf{G}^{(t)}$ to be a diagonal matrix for each t. Then, we simply have

$$\mathbf{G}^{(t)} = \text{diag}\left(\sum_{i=0}^{t} \nabla f(\mathbf{w}^{(i)}) \odot \nabla f(\mathbf{w}^{(i)})\right), \tag{3.36}$$

$$\mathbf{w}^{(t+1)} = \mathbf{w}^{(t)} - \eta(\mathbf{G}^{(t)})^{-1/2}\nabla f(\mathbf{w}^{(t)}), \tag{3.37}$$

where $\odot$ denotes pointwise multiplication. Essentially, both (3.36) and (3.37) only contain pointwise operations and can be implemented efficiently. In particular, we can perform pointwise division in (3.37) and need not really use the diagonal matrices. That is, we can simply implement the following:

$$\mathbf{g}^{(t)} = \left(\sum_{i=0}^{t} \nabla f(\mathbf{w}^{(i)}) \odot \nabla f(\mathbf{w}^{(i)}) \right), \tag{3.38}$$

$$\mathbf{w}^{(t+1)} = \mathbf{w}^{(t)} - \eta \nabla f(\mathbf{w}^{(t)}) \oslash \sqrt{\mathbf{g}^{(t)}}, \tag{3.39}$$

where $\oslash$ denotes pointwise division and the square root is understood as a pointwise operation in (3.39). For better numerical stability, it is custom to use $\sqrt{\mathbf{g}^{(t)} + \epsilon}$, with a fixed small number ϵ (say 10^{-10}), in place of $\sqrt{\mathbf{g}^{(t)}}$.

Very similar to AdaGrad, but probably more widely used, is the update scheme called *RMSProp*, or "Root Mean Squared Propagation", which originally appeared in 2014, in a lecture on neural networks given by Geoffrey Hinton. Note that "Root Mean Squared" only describes what has been done in (3.36). The main difference between RMSProp and AdaGrad is that RMSProp uses an exponentially weighted moving average of $\mathbf{G}^{(t)}$ in the gradient step. Specifically, with $\mathbf{R}^{(0)}$ initialized to be a zero matrix, the update steps of RMSProp are as follows:

$$\mathbf{G}^{(t)} = \operatorname{diag} \left(\sum_{i=0}^{t} \nabla f(\mathbf{w}^{(i)}) \odot \nabla f(\mathbf{w}^{(i)}) \right), \tag{3.40}$$

$$\mathbf{R}^{(t)} = \alpha \mathbf{R}^{(t-1)} + (1 - \alpha) \mathbf{G}^{(t)}, \tag{3.41}$$

$$\mathbf{w}^{(t+1)} = \mathbf{w}^{(t)} - \eta (\mathbf{R}^{(t)})^{-1/2} \nabla f(\mathbf{w}^{(t)}). \tag{3.42}$$

It is also custom to simply use $\nabla f(\mathbf{w}^{(t)}) \odot \nabla f(\mathbf{w}^{(t)})$ directly in (3.40), instead of the sum over all i from 0 to t. Similar to AdaGrad, (3.40)–(3.42) are essentially pointwise operations and can be implemented efficiently.

Problem 3.3. Write the pointwise steps of (3.40)–(3.42), similar to (3.38)–(3.39).

Another widely used optimization method is *Adam*, originally proposed by Kingma and Ba (2015). The main difference between AdaGrad/RMSProp and Adam is that Adam adopts the momentum regime. The update steps of Adam are as follows:

$$\mathbf{g}^{(t)} = \nabla f(\mathbf{w}^{(t)}), \tag{3.43}$$

$$\mathbf{m}^{(t)} = \beta_1 \mathbf{m}^{(t-1)} + (1 - \beta_1)\mathbf{g}^{(t)}, \tag{3.44}$$

$$\mathbf{s}^{(t)} = \beta_2 \mathbf{s}^{(t-1)} + (1 - \beta_2)\mathbf{g}^{(t)} \odot \mathbf{g}^{(t)}, \tag{3.45}$$

$$\hat{\mathbf{m}}^{(t)} = \mathbf{m}^{(t)}/(1 - \beta_1^t), \tag{3.46}$$

$$\hat{\mathbf{s}}^{(t)} = \mathbf{s}^{(t)}/(1 - \beta_2^t), \tag{3.47}$$

$$\mathbf{w}^{(t+1)} = \mathbf{w}^{(t)} - \eta\hat{\mathbf{m}}^{(t)} \oslash \sqrt{\hat{\mathbf{s}}^{(t)}}, \tag{3.48}$$

where again $\oslash$ denotes pointwise division and the square root is understood as a pointwise operation.

To conclude this section, we list some examples of optimizers implemented in PyTorch, using AdaGrad, RMSProp, and Adam, respectively.

```
opt_Adagrad = torch.optim.Adagrad(model.parameters(),lr=1e-3)
opt_RMSprop =
    torch.optim.RMSprop(model.parameters(),lr=1e-3,alpha=0.9)
opt_Adam =
    torch.optim.Adam(model.parameters(),lr=1e-3,betas=(0.9,0.99))
```

3.3.1 *Analysis of convergence for AdaGrad**

Next, let's perform a quick analysis of the convergence of AdaGrad with (3.34) and (3.35). Let's assume that f is convex as before. For a positive semi-definite matrix $\mathbf{A}$, define $\|\mathbf{w}\|_{\mathbf{A}}^2 = \mathbf{w}^\top \mathbf{A}\mathbf{w}$. We have

$$\|\mathbf{w}^{(t+1)} - \mathbf{w}^*\|_{\mathbf{G}^{(t)}}^2 \tag{3.49}$$

$$= \|\mathbf{w}^{(t)} - \eta(\mathbf{G}^{(t)})^{-1}\nabla f(\mathbf{w}^{(t)}) - \mathbf{w}^*\|_{\mathbf{G}^{(t)}}^2 \tag{3.50}$$

$$= \|\mathbf{w}^{(t)} - \mathbf{w}^*\|_{\mathbf{G}^{(t)}}^2 + \eta^2 \|\nabla f(\mathbf{w}^{(t)})\|_{(\mathbf{G}^{(t)})^{-1}}^2$$

$$- 2\eta\nabla f(\mathbf{w}^{(t)})^\top(\mathbf{w}^{(t)} - \mathbf{w}^*), \tag{3.51}$$

which can be rearranged as

$$\nabla f(\mathbf{w}^{(t)})^\top (\mathbf{w}^{(t)} - \mathbf{w}^*)$$

$$= \frac{1}{2\eta}(\|\mathbf{w}^{(t)} - \mathbf{w}^*\|_{\mathbf{G}^{(t)}}^2 - \|\mathbf{w}^{(t+1)} - \mathbf{w}^*\|_{\mathbf{G}^{(t)}}^2)$$

$$+ \frac{\eta}{2}\|\nabla f(\mathbf{w}^{(t)})\|_{(\mathbf{G}^{(t)})^{-1}}^2. \tag{3.52}$$

Since f is convex, $f(\mathbf{w}^{(t)}) - f(\mathbf{w}^*) \leq \nabla f(\mathbf{w}^{(t)})^\top (\mathbf{w}^{(t)} - \mathbf{w}^*)$. Summing from $t = 0$ to $t = T - 1$, we have

$$\sum_{t=0}^{T-1}(f(\mathbf{w}^{(t)}) - f(\mathbf{w}^*)) \tag{3.53}$$

$$\leq \frac{1}{2\eta}\sum_{t=1}^{T-1}(\|\mathbf{w}^{(t)} - \mathbf{w}^*\|_{\mathbf{G}^{(t)}}^2 - \|\mathbf{w}^{(t)} - \mathbf{w}^*\|_{\mathbf{G}^{(t-1)}}^2) \tag{3.54}$$

$$+ \frac{1}{2\eta}\|\mathbf{w}^{(0)} - \mathbf{w}^*\|_{\mathbf{G}^{(0)}}^2 - \frac{1}{2\eta}\|\mathbf{w}^{(T)} - \mathbf{w}^*\|_{\mathbf{G}^{(T-1)}}^2 \tag{3.55}$$

$$+ \frac{\eta}{2}\sum_{t=0}^{T-1}\|\nabla f(\mathbf{w}^{(t)})\|_{(\mathbf{G}^{(t)})^{-1}}^2. \tag{3.56}$$

Next, we need to have estimates of the above three lines. Let's assume we are searching for $\mathbf{w}^*$ within a ball centered at $\mathbf{w}^*$ with radius D. Then,

$$\sum_{t=1}^{T-1}(\|\mathbf{w}^{(t)} - \mathbf{w}^*\|_{\mathbf{G}^{(t)}}^2 - \|\mathbf{w}^{(t)} - \mathbf{w}^*\|_{\mathbf{G}^{(t-1)}}^2) \tag{3.57}$$

$$= \sum_{t=1}^{T-1}(\mathbf{w}^{(t)} - \mathbf{w}^*)^\top (\mathbf{G}^{(t)} - \mathbf{G}^{(t-1)})(\mathbf{w}^{(t)} - \mathbf{w}^*) \tag{3.58}$$

$$\leq \sum_{t=1}^{T-1}\lambda_{\max}(\mathbf{G}^{(t)} - \mathbf{G}^{(t-1)})D^2 \tag{3.59}$$

where $\lambda_{\max}(\cdot)$ denotes the largest eigenvalue

$$\leq \sum_{t=1}^{T-1} \operatorname{tr}(\mathbf{G}^{(t)} - \mathbf{G}^{(t-1)})D^2 \tag{3.60}$$

where $\operatorname{tr}(\cdot)$ denotes the trace of a matrix

$$= \sum_{t=1}^{T-1} (\operatorname{tr}(\mathbf{G}^{(t)}) - \operatorname{tr}(\mathbf{G}^{(t-1)}))D^2 \tag{3.61}$$

$$\leq \operatorname{tr}(\mathbf{G}^{(T)})D^2. \tag{3.62}$$

Similarly,

$$\|\mathbf{w}^{(0)} - \mathbf{w}^*\|_{\mathbf{G}^{(0)}}^2 \leq \operatorname{tr}(\mathbf{G}^{(0)})D^2. \tag{3.63}$$

Also, we leave it as an exercise to show the following.

Problem 3.4. Show that $\sum_{t=0}^{T-1} \|\nabla f(\mathbf{w}^{(t)})\|_{(\mathbf{G}^{(t)})^{-1}}^2 \leq 2\operatorname{tr}(\mathbf{G}^{(T)})$.

Altogether, we have

$$\sum_{t=0}^{T-1} (f(\mathbf{w}^{(t)}) - f(\mathbf{w}^*)) \leq \frac{1}{2\eta}\operatorname{tr}(\mathbf{G}^{(T)})D^2 + \frac{\eta}{2}2\operatorname{tr}(\mathbf{G}^{(T)}). \tag{3.64}$$

Taking $\eta = D$, we have

$$\sum_{t=0}^{T-1} (f(\mathbf{w}^{(t)}) - f(\mathbf{w}^*)) \leq \frac{3D}{2}\operatorname{tr}(\mathbf{G}^{(T)}), \tag{3.65}$$

or equivalently,

$$\frac{1}{T}\sum_{t=0}^{T-1} (f(\mathbf{w}^{(t)}) - f(\mathbf{w}^*)) \leq \frac{1}{T}\frac{3D}{2}\operatorname{tr}(\mathbf{G}^{(T)}). \tag{3.66}$$

Therefore, $f(\mathbf{w}^{(T)}) - f(\mathbf{w}^*) = O(1/T)$ and we have concluded the convergence analysis for AdaGrad.

3.4 Initialization

Good initialization plays a crucial role in iterative optimization methods, especially in training neural networks. The complexity of the loss landscape in neural networks requires careful initialization to ensure both convergence and efficiency. In particular, the issue of vanishing

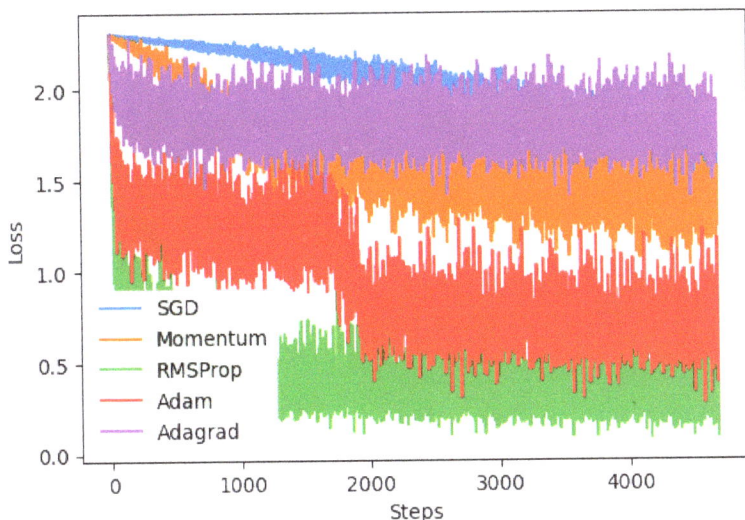

Fig. 3.2. An empirical comparison among optimization methods.

or exploding gradients during gradient propagation highlights the importance of a proper initialization strategy.

To address this challenge, a popular heuristic for weight initialization in dense networks is to keep the variance of each layer the same after applying activation functions. For instance, one popular initialization method is the Xavier normal initialization, where each weight is initialized randomly according to the Gaussian distribution $\mathcal{N}(0, \sigma^2)$, where

$$\sigma^2 = \text{gain} \times \frac{2}{\#\text{in} + \#\text{out}}.$$

Here, "#in" is the number of input neurons in the current layer and "#out" is the number of output neurons; "gain" is a scalar normalization factor which takes into account the activation function used (a list of gain values can be found at, e.g., https://pytorch.org/docs/stable/nn.init.html).

We end this chapter with the following numerical exercise.

Problem 3.5. To get some feeling about performance of difference optimization methods, try to implement SGD, SGD with momentum, RMSProp, Adam, and AdaGrad for the classification problem in Chapter 1. Plot the training loss versus the number of iteration steps. A sample result is plotted in Figure 3.2.

PART 2
Representation Learning

Chapter 4

Autoencoders and Principal Components Analysis

4.1 Autoencoder

Representation Learning is closely related to the desire to capture and possibly visualize the structure of data in a way that is intuitive, meaningful, and compact.

Let's explore the concept of representation learning using a coffee analogy. Imagine we have different types of coffee drinks with names that may not immediately reveal their ingredients, such as Americano, Latte, and Mocha. To better understand and represent these coffee drinks, we can define a basis of coffee components, such as espresso, hot water, steamed milk, milk foam, hot chocolate, cream, ice cream, and whiskey. Using this basis, we can express each coffee drink as a unique combination of these components. For example, an Americano can be represented as a mixture of one hot water and one espresso, denoted as $(1, 1, 0, 0, 0, 0, 0, 0)$. A Latte, on the other hand, contains one espresso, one steamed milk, and one milk foam, represented as $(1, 0, 1, 1, 0, 0, 0, 0)$. Similarly, a Mocha can be represented as $(1, 0, 1, 1, 1, 0, 0, 0)$. In this way, each coffee drink can be understood and represented as a unique combination of the underlying coffee components. This representation simplifies the understanding and categorization of different types of coffee, even for drinks with less intuitive names. The process of determining these representations from the basis of coffee components is analogous to how representation learning in machine learning automatically discovers informative features from raw data.

As such, methods of dimensionality reduction are natural candidates for accomplishing the representation learning task. Another important aspect of learning representation is that the problem is often defined in "unsupervised" manner. Although one could ask the question of what is a "useful" or compact or meaningful representation of your data for a supervised task or, in other words, find ways to efficiently parse or cluster the data of source X when a target Y is provided, in many data science tasks, the amount of labeled data is small or the cost of labeling is prohibitively high. Moreover, in some cases, the labeling itself is problematic because the categories are ill-defined. If the task is simply compression, visualization, or data exploration, then no labeling is available. In such situations, we might want to find a transformation of the data into a another representation that has some optimal properties. For this purpose, representation learning deals with finding a set of basis functions that are fitted by solving some optimization task. After learning is accomplished, these pre-existing basis functions can be further used to transform new data into the new target coordinates. The basis function later can be used to fit another regression model or initialize a neural network and only train for a short time. In fact, unsupervised "pre-training" was one of the genius solutions that led to the revival, and today's explosion, in deep learning.

In this section, we build on the notion of the autoencoder, also commonly spelled as Auto-Encoder (AE), as a way to define the representation learning task. AE's task is to find a representation of feature vectors without any label that can be later used to reconstruct the original data. Moreover, AE is defined as a two-step mapping or function from data to features and from features back to data. As such, one common way to denote the two-step nature is by calling the first part of going from data to features as "encoder" and then back to data as "decoder". This immediately suggests a relation to information transmission, where the tasks of the encoding are to find a compact representation. It should be noted of course that compression is not the only viable criteria for AE, but the common idea is that there is some sort of a bottleneck that forces the data to get rid of its irrelevant aspects, leaving only the essential aspects that are needed for later reconstruction. The encoder function can be viewed as the transformation of the original data into another space, where the values of the feature vector found at the output of the encoder are considered as weights of the basis function of that space. Formally, this can be implemented as matrix transformation, neural network, and more.

For the purpose of this chapter, it would be useful to consider matrix transformation as a particular case of a neural network without activation functions (or with identity activation functions). Later on, we also show how the same problem can be solved as a closed-form inverse matrix problem in the case where the neurons are linear (or in other words, when there is no nonlinear activation function present in mapping the weighted inputs to outputs).

An AE is a vector-valued function that is trained to approximately return its input $\mathbf{x} \approx \mathbf{F}(\mathbf{x})$. In the simplest case, we set up a transformation matrix $\mathbf{T}$ to define our mapping as $\mathbf{F}(\mathbf{x}) = \mathbf{T}\mathbf{x}$, where we also may find the trivial solution with $\mathbf{T} = \mathbf{I}$ as identity mapping. Accordingly, some constraints are required to find an interesting representation, such as reducing the dimension of some internal step.

4.2 Dimensionality Reduction

One possible constraint is to form a bottleneck by the use of a neural network with input of dimension N and hidden layer of K units where $K \ll N$. Specifically, we apply functions $\mathbf{h}$ and $\mathbf{f}$ such that

$$\mathbf{h}(\mathbf{x}) = g(\mathbf{V}\mathbf{x} + \mathbf{b}^{(1)}), \tag{4.1}$$

$$\mathbf{f}(\mathbf{h}(\mathbf{x})) = g(\mathbf{W}\mathbf{h}(\mathbf{x}) + \mathbf{b}^{(2)}), \tag{4.2}$$

where $\mathbf{V}$ is a $K \times N$ weight matrix, $\mathbf{W}$ is a $N \times K$ weight matrix, and the g's are element-wise nonlinear activation functions. See Fig. 4.1 for an illustration of these neural networks. The function pair $\mathbf{h}, \mathbf{f}$ can be considered as encoder and decoder for a lossy compressor. The network compresses an input vector of N values down into a smaller set of K numbers and then decodes them again back into the original dimension, approximately reconstructing the original input. This creates effectively a mapping $\mathbf{F}(\mathbf{x}) = \mathbf{f}(\mathbf{h}(\mathbf{x})) \approx \mathbf{x}$, where the quality of approximation or the reconstruction error needs yet to be defined.

$$x \in \mathbb{R}^N \qquad\qquad\qquad F(x) \in \mathbb{R}^N$$

$$h(x) \in \mathbb{R}^K$$

$$h(x) = g(Vx + b^{(1)}) \qquad\qquad F(x) = g(Wh(x) + b^{(2)})$$

$$\underbrace{\phantom{h(x) = g(Vx + b^{(1)})}}_{\text{encoder}} \qquad\qquad \underbrace{\phantom{F(x) = g(Wh(x) + b^{(2)})}}_{\text{decoder}}$$

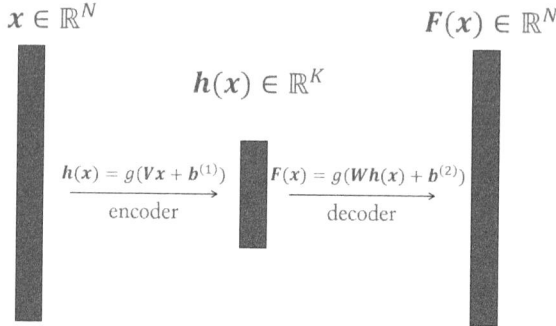

Fig. 4.1. The AE defined by (4.1) and (4.2).

4.3 Principal Components Analysis (PCA)

We consider the Principal Components Analysis (PCA) as a linear autoencoder where activation function is an identity function $g(a) = a$:

$$\mathbf{h} = \mathbf{V}\mathbf{x} + \mathbf{b}^{(1)}, \qquad (4.3)$$

$$\mathbf{f} = \mathbf{W}\mathbf{h} + \mathbf{b}^{(2)}. \qquad (4.4)$$

Without loss of generality, we may assume that the mean of our data $\mathbf{x}$ is zero, which can be shown to allow removing the bias terms $\mathbf{b}^{(1)}$ and $\mathbf{b}^{(2)}$. The goal of our minimization problem now becomes minimizing an error function between the network output $\mathbf{F}(\mathbf{x}) = \mathbf{W}\mathbf{V}\mathbf{x}$ and the desired output $\mathbf{y}$, which in the autoencoding case is $\mathbf{x}$ itself. Without loss of generality, let us write the error for any target $\mathbf{y}$ as

$$E(\mathbf{V}, \mathbf{W}) = \|\mathbf{y} - \mathbf{W}\mathbf{V}\mathbf{x}\|^2 = (\mathbf{y} - \mathbf{W}\mathbf{V}\mathbf{x})^\top (\mathbf{y} - \mathbf{W}\mathbf{V}\mathbf{x}). \qquad (4.5)$$

The goal of our neural network is to learn the parameters of the matrices $\mathbf{V}$ and $\mathbf{W}$ to minimize the error E, which eventually will lead to saddle or extreme point where $\frac{\partial E}{\partial v_{ij}} = 0$ and $\frac{\partial E}{\partial w_{ij}} = 0$, where v_{ij} and w_{ij} are elements of the matrices $\mathbf{V}$ and $\mathbf{W}$, respectively.

Let us denote the output of the first step of the transformation as $\mathbf{z}$, i.e., $\mathbf{z} = \mathbf{h}(\mathbf{x})$. This vector is indeed the compact representation or the latent or hidden layer in the encoder–decoder network, also sometimes referred as "features". At this point, we have not yet specified what the constraints are on the behavior of $\mathbf{z}$ apart for requiring it to be of a lower dimension $K < N$. Throughout this book, we revisit this same formulation by adding other constraints on the desired features.

Our task is to show that a neural network that learns through the process of iterative parameter optimization, such as backpropagation, indeed arrives for the autoencoding case, to the classical solution of PCA, which is finding the eigenvectors of the data correlation matrix $\Sigma_{\mathbf{XX}}$ and using the first K eigenvectors with the largest eigenvalues as the columns of the matrix $\mathbf{V}$ or rows of $\mathbf{W}$. For those not familiar or needing refreshing of PCA, the PCA algorithm can be summarized as the following Python code:

```python
# Find top K principal directions:
# Input data X contains multiple rows of samples x, each of
    dimension N
L,V = np.linalg.eig(np.cov(X.T))
iL = np.argsort(L)
V = V[:, iL[:K]] # reduction of N to K
x_bar = np.mean(X, 0)
# Reduce X to K dimensions
Z = np.dot(X - x_bar, V) # (K,N)
# Transform back to N dimensions:
W = V.T
X_proj = np.dot(Z, W) + x_bar # (N,K)
```

Figure 4.2 illustrates an example of PCA, where data points reside in a three-dimensional space and PCA describes a two-dimensional subspace that captures the main variance of the data points.

The goal of the remaining of the chapter is to show that a neural network with linear neurons achieves the same solution. This can be also viewed as an alternative proof of the PCA principle, i.e., matrices V and W that PCA finds are a solution of the optimization problem and thus they are the best representation in terms of the mean squared error (MSE) loss for a data reduction to a lower dimensional space.

Let us assume first that we have a fixed decoding matrix $\mathbf{W}$. Finding the least error E for such a case amounts to finding the best set of features $\mathbf{z}$ so that $E(\mathbf{z}) = \|\mathbf{y} - \mathbf{Wz}\|^2 = \|\mathbf{y}\|^2 - 2\mathbf{y}^\top \mathbf{Wz} + \mathbf{z}^\top \mathbf{W}^\top \mathbf{Wz}$ is minimized. Taking the derivative, one finds the solution $\mathbf{z} = (\mathbf{W}^\top \mathbf{W})^{-1} \mathbf{W}^\top \mathbf{y}$.

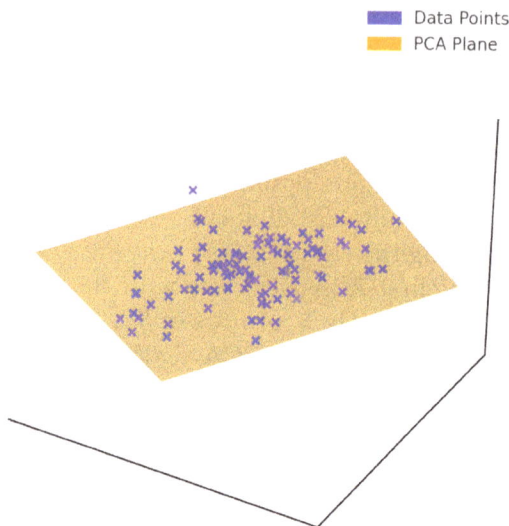

Fig. 4.2.   An example of PCA that finds a two-dimensional subspace for three-dimensional data points.

The expression $\text{pInv}(\mathbf{W}) := (\mathbf{W}^\top\mathbf{W})^{-1}\mathbf{W}^\top$ is known as the pseudo-inverse of $\mathbf{W}$. In other words, if we have a transformation that reduces the dimension of $\mathbf{y}$ into K-dimensional vector $\mathbf{z}$ by the relation $\mathbf{y} \approx \mathbf{W}\mathbf{z}$, the best $\mathbf{z}$ for a fixed $\mathbf{W}$ is $\mathbf{z} = \text{pInv}(\mathbf{W})\mathbf{y} := (\mathbf{W}^\top\mathbf{W})^{-1}\mathbf{W}^\top\mathbf{y}$. The total transformation of the input $\mathbf{x}$ to output $\mathbf{y}$ can now be written as $\mathbf{y} = \mathbf{W}\mathbf{V}\mathbf{x}$.

Going back to the autoencoding case $\mathbf{y} = \mathbf{x}$, we see that best features to represent $\mathbf{x}$ are $\mathbf{z} = \mathbf{V}\mathbf{x} = (\mathbf{W}^\top\mathbf{W})^{-1}\mathbf{W}^\top\mathbf{x}$, so the encoding matrix $\mathbf{V}$ should be the pInv of $\mathbf{W}$. The total transformation of $\mathbf{x}$ through the network becomes $\mathbf{W}\mathbf{V}\mathbf{x} = \mathbf{W}(\mathbf{W}^\top\mathbf{W})^{-1}\mathbf{W}^\top\mathbf{x} = \mathbf{P}_\mathbf{W}\mathbf{x}$, where $\mathbf{P}_\mathbf{W}$ is known as a projection matrix $\mathbf{P}_\mathbf{W} = \mathbf{W}(\mathbf{W}^\top\mathbf{W})^{-1}\mathbf{W}^\top$ that projects any vector of dimension N into a lower-dimensional space K spanned by the columns of the matrix $\mathbf{W}$ (in our notation, $\mathbf{W}$ is (N, K)).

Returning to our error function, we can write now for the average error for an autoencoding case

$$E(\mathbf{V}, \mathbf{W}) = \langle \|\mathbf{x}\|^2 - 2\mathbf{x}^\top\mathbf{P}_\mathbf{W}\mathbf{x} + \mathbf{x}^\top\mathbf{P}_\mathbf{W}^2\mathbf{x}\rangle$$
$$= \langle\|\mathbf{x}\|^2\rangle - \langle\mathbf{x}^\top\mathbf{P}_\mathbf{W}\mathbf{x}\rangle = \text{tr}(\mathbf{\Sigma}_{\mathbf{XX}}) - \text{tr}(\mathbf{P}_\mathbf{W}\mathbf{\Sigma}_{\mathbf{XX}}), \tag{4.6}$$

where we used the relation $\mathbf{P}_\mathbf{W}^2 = \mathbf{P}_\mathbf{W}$ (projecting the results of a projection on more time does not change it), and the multiplication property of a trace of a matrix $\text{tr}(\mathbf{A}^\top \mathbf{B}) = \text{tr}(\mathbf{B}\mathbf{A}^\top)$, and used the notation $\Sigma_{\mathbf{XX}}$ to denote a correlation matrix of $\mathbf{X}$ that can be written as $\Sigma_{\mathbf{XX}} = \frac{1}{M}\mathbf{X}^\top \mathbf{X}$, where M is the number of samples, with a slight abuse of notation, since we introduced here the averaging of multiple samples of $\mathbf{x}$ arranged into a matrix $\mathbf{X}$ as approximation to statistical mean, which allowed us to consider now the average error as a function of the correlation matrix.

This gives us the following insight into how to choose the K basis vectors

$$E(\mathbf{V}, \mathbf{W}) = \text{tr}(\Sigma_{\mathbf{XX}}) - \text{tr}(\mathbf{P}_{\mathbf{W}}\Sigma_{\mathbf{XX}}) = \sum_{i=1}^{N} \lambda_i - \sum_{i \in I} \lambda_i, \qquad (4.7)$$

or in other words, the error becomes the sum of the residual eigenvalues for the eigenvectors we did not choose to go into our projection.

We are almost done here, so before proceeding to the last PCA step, let us discuss the expression above. What we see it that the optimal encoding and decoding over a large set of data points can be achieved by finding an encoding matrix $\mathbf{W}$ so that the difference between the trace of the data correlation matrix and the trace of the correlation matrix projected into a lower-dimensional space spanned by the encoder is minimized. We are ready now to use couple more neat mathematical properties of correlation matrices to complete our proof of optimality of PCA.

First, we want to explore the eigenvectors of $\Sigma_{\mathbf{XX}}$. Let us denote by $\mathbf{U}$ a matrix of its eigenvectors (N vectors of N dimensions) and by Λ the diagonal matrix of its eigenvalues. The expression for eigenvector relations is by definition $\Sigma_{\mathbf{XX}}\mathbf{U} = \mathbf{U}\Lambda$. Since $\Sigma_{\mathbf{XX}}$ is real symmetric matrix, its eigenvectors are a complete orthonormal set, i.e., the vectors $\mathbf{u}_i$, $\mathbf{u}_j$ for any member of the set of eigenvectors $\mathbf{U} = [\mathbf{u}_1, \mathbf{u}_2, \dots, \mathbf{u}_N]$ are orthonormal, i.e., $\mathbf{u}_i^\top \mathbf{u}_j = \delta_{i,j}$, or zero when $i \neq j$ and one when $i = j$.

Second, let us choose encoding matrix $\mathbf{W}$ to be a subset of K vectors from $\mathbf{U}$. The only remaining task would be to decide which K vectors to choose. Let us denote this subset by index I, so $\mathbf{W} = \mathbf{U}_I$. Going back to the expression of the error $E(\mathbf{V}, \mathbf{W})$, we can write $\mathbf{P}_{\mathbf{W}} = \mathbf{U}_I(\mathbf{U}_I^\top \mathbf{U})^{-1}\mathbf{U}_I^\top = \mathbf{U}_I\mathbf{U}_I^\top$ where we eliminated the middle inverse part since $\mathbf{U}_I^\top \mathbf{U}_I = \mathbf{I}$ because of its orthonormality.

We leave this as a proof for the reader to show that for any other matrix that is not aligned with the eigenvectors $\mathbf{U}_I$, the trace of

$\mathbf{P_W} \Sigma_{\mathbf{XX}}$ will be smaller than $\sum_{i \in I} \lambda_i$. Intuitively, this can be seen as generalization of a projection in a one-dimensional case. It is clear that if we want to project a given vector on some basis function, the largest projection will be in the direction of the vector. In other words, if we can rotate the projection vector, the projection will be zero for an orthogonal vector and then proportional to the cosine of the angle, which will be maximal when the angle is zero.

We thus have demonstrated that choosing K eigenvectors of $\Sigma_{\mathbf{XX}}$ that have the largest eigenvalues gives the best encoding of the data $\mathbf{X}$ into a lower K-dimensional space, with the remaining error being the sum of the eigenvalues of the "missed" dimensions.

In practice, this situation is not guaranteed in neural network types of optimization. The process of finding the encoding matrix parameters can be stuck in local extrema (despite the fact that the problem is convex and has a single global minimum).

The advantages of finding a closed form solution of PCA instead of using an autoencoder with gradient methods are as follows:

- There are efficient solvers for finding eigenvectors that might be faster than gradient search.
- With enough data and sufficiently good estimate of the data covariance, we know that this is an optimal solution.
- We have better intuition into the meaning of the latent representation (features) while NN remains a black box.

On the other hand, there is a big deficiency with PCA compared to NN autoencoder. By allowing a nonlinear function $g(\cdot)$ to process the encoding and decoding vectors, we can successively apply autoencoding in multiple steps, each one gradually reducing the dimensionality and the amount of information extracted from the data. Such gradual approach is not possible in the linear case (no activation function), since any product of matrices can be represented as a single final matrix, so gradual reduction of dimensions from N to $K_1 < N$ to $K_2 < K_1$ is equivalent to reducing it straight to K_2.

4.4   Denoising AE

When the units of AE are linear with architecture that has a bottleneck shape (i.e., the dimension of hidden units is smaller than the dimension of the data), and the loss is a mean squared error, the representation it finds is similar to that of PCA. One way to think about it is that minimization of the error during learning drives the AE to seek the best representation, which consequently forces the connection matrix to reside in a subspace that is spanned by the principal components of the data.

Autoencoders with nonlinear activation of its neurons are able to learn much more powerful representation. One can think about it as a nonlinear generalization of PCA. Moreover, the nonlinearity allows gradual reduction of representation complexity through multiple layers, allowing for very complex function mapping from input to the hidden units and to the output. Unfortunately, such powerful mapping may result in AE that learns to memorize the training set, mapping input to out without extracting useful information about its distribution. There are several ways to try to avoid this over-fitting problem. One of them is the variational autoencoder (VAE) that is discussed in later chapters. Before going into such modeling, it makes sense to think about this problem through a regularization approach. In regularization, rather than limiting the model capacity by keeping the encoder and decoder shallow and the hidden state dimension small, the solution is to add an additional component to the loss function that forces the model to have some smooth or constrained hidden layer properties that will avoid copying input to output. Another solution to the exact recopying is even more ingenious — instead of feeding the same data in the input and output and expecting the AE to learn the best possible reconstructed, the denoising AE feeds a corrupted version of the data into the input and cleans the data at the output. The denoising autoencoder (DAE) is trained to predict the original uncorrupted data point as its output.

Such prediction can be thought of as a map from corrupted data $\hat{\mathbf{x}}$ that is distributed according to probability $\hat{\mathbf{x}} \sim C(\hat{\mathbf{x}}|\mathbf{x})$, pointing it back to the original data point $\mathbf{x}$. If we assume that $\mathbf{x}$ resides on a smooth lower-dimensional surface (manifold), adding noise most likely will throw the data in directions that are orthogonal to the manifold. Of course, we do not know what the manifold is, but we assume that it has a lower dimension than the data itself. So, for large dimensional data residing in a low-dimensional space, adding noise will locally effect only few dimensions, while the majority of the noise will happen in other dimensions. When the denoising autoencoder is trained to minimize the average of squared errors $\|g(f(\hat{\mathbf{x}})) - \mathbf{x}\|^2$, with f and g denoting the encoder and decoder, respectively, it can be shown that the reconstruction $g(f(\hat{\mathbf{x}}))$ esti-mates $\mathbb{E}_{(\mathbf{x},\hat{\mathbf{x}})\sim p_{\text{data}}(\mathbf{x})C(\hat{\mathbf{x}}|\mathbf{x})}[\mathbf{x}|\hat{\mathbf{x}}]$, which is the center of mass of the clean points $\mathbf{x}$ that could have resulted in samples $\hat{\mathbf{x}}$ after applica-tion of the corruption. The difference vector $g(f(\hat{\mathbf{x}})) \sim \hat{\mathbf{x}}$ does an approximate projection on the clear signal manifold.

4.5   A PyTorch Implementation of AE

Let's use PyTorch to build an AE for the Fashion MNIST dataset. First, we define a class for encoder and a class for decoder:

```
import torch
import torch.nn as nn
import torch.nn.functional as F
import torch.optim as optim
import torchvision.transforms as transforms
from torchvision.datasets import FashionMNIST
from torch.utils.data import DataLoader

device = 'cuda' if torch.cuda.is_available() else 'cpu'

class Encoder(nn.Module):
    def __init__(self, latent_dims):
        super(Encoder, self).__init__()
        self.linear1 = nn.Linear(784, 512)
        self.batch_norm1 = nn.BatchNorm1d(512)
        self.linear2 = nn.Linear(512, latent_dims)

    def forward(self, x):
        x = torch.flatten(x, start_dim=1)
        x = F.relu(self.batch_norm1(self.linear1(x)))
        return self.linear2(x)

class Decoder(nn.Module):
    def __init__(self, latent_dims):
        super(Decoder, self).__init__()
        self.linear1 = nn.Linear(latent_dims, 512)
        self.batch_norm1 = nn.BatchNorm1d(512)
        self.linear2 = nn.Linear(512, 784)

    def forward(self, z):
        z = F.relu(self.batch_norm1(self.linear1(z)))
        z = torch.sigmoid(self.linear2(z))
        return z.view(-1, 1, 28, 28)
```

An AE is simply a cascading of the encoder and the decoder:

```python
class Autoencoder(nn.Module):
    def __init__(self, latent_dims):
        super(Autoencoder, self).__init__()
        self.encoder = Encoder(latent_dims)
        self.decoder = Decoder(latent_dims)

    def forward(self, x):
        z = self.encoder(x)
        return self.decoder(z)
```

We can write a training function as follows.

```python
def train(autoencoder, data, epochs=5):
    opt = torch.optim.Adam(autoencoder.parameters())
    scheduler = torch.optim.lr_scheduler.StepLR(opt, step_size=10,
        gamma=0.1) # This is a scheduler we have not discussed.
        Check it!
    for epoch in range(epochs):
        total_loss = 0
        for x, _ in data:
            x = x.to(device)
            opt.zero_grad()
            x_hat = autoencoder(x)
            loss = F.mse_loss(x_hat, x)
            loss.backward()
            opt.step()
            total_loss += loss.item()
        scheduler.step()
        print(f"Epoch {epoch+1}, Loss: {total_loss/len(data)}")
    return autoencoder
```

Now, we are ready to train the AE using Fashion MNIST!

```python
transform = transforms.Compose([transforms.ToTensor()])
train_dataset = FashionMNIST(root='./data', train=True,
    download=True, transform=transform)
train_loader = DataLoader(train_dataset, batch_size=64,
    shuffle=True)

latent_dims = 2
autoencoder = Autoencoder(latent_dims).to(device)
trained_model = train(autoencoder, train_loader)
```

Here, we have set the latent dimension to be 2. It is interesting to look at the reconstructed images. Actually, we can choose our latent code **z** and see how the reconstructed images change with it:

```python
import numpy as np
import matplotlib.pyplot as plt

def plot_reconstructed(autoencoder, r0=(-5, 10), r1=(-10, 5), n=12):
    autoencoder.eval() # This is needed since we have used batch
        normalization in the autoencoder!
    w = 28
    img = np.zeros((n*w, n*w))
    for i, y in enumerate(np.linspace(*r1, n)):
        for j, x in enumerate(np.linspace(*r0, n)):
            z = torch.Tensor([[x, y]]).to(device)
            x_hat = autoencoder.decoder(z)
            x_hat = x_hat.reshape(28, 28).cpu().detach().numpy()
            img[(n-1-i)*w:(n-1-i+1)*w, j*w:(j+1)*w] = x_hat
    plt.imshow(img, extent=[*r0, *r1], cmap='gray')
    plt.title("Reconstructed Images from Latent Space")
    plt.show()

plot_reconstructed(trained_model)
```

A sample output is illustrated in Figure 4.3.

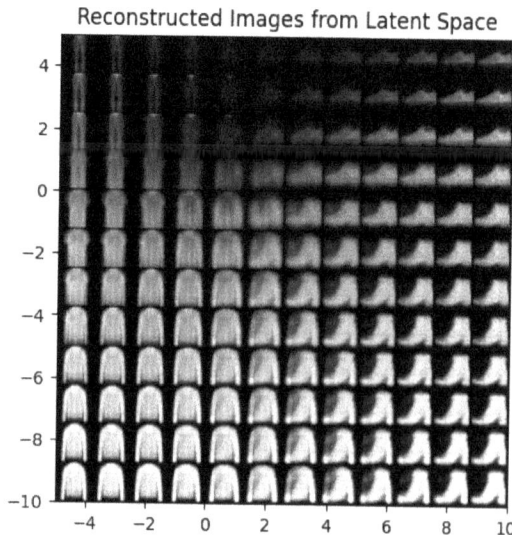

Fig. 4.3.   Reconstructed images using AE with a two-dimensional latent space.

We can also illustrate the latent space of the training data using the following code:

```python
def plot_latent(autoencoder, dataloader, num_batches=50):
    autoencoder.eval() # Set the model to evaluation mode
    fig, ax = plt.subplots()
    for i, (x, y) in enumerate(dataloader):
        with torch.no_grad():
            z = autoencoder.encoder(x.to(device))
        z = z.to('cpu').numpy()
        y = y.numpy() # Assuming y is a tensor, need to convert to
            numpy
        scatter = ax.scatter(z[:, 0], z[:, 1], c=y, cmap='tab10',
    alpha=0.6)

        if i >= num_batches - 1: # Corrected to handle exactly
            num_batches
            break

    plt.colorbar(scatter, ax=ax)
    plt.title("Latent Space Visualization")
    plt.show()

plot_latent(autoencoder, train_loader)
```

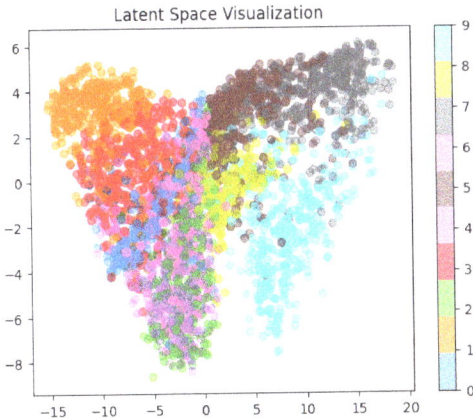

Fig. 4.4. Reconstructed images using AE with a two-dimensional latent space.

A sample output is illustrated in Figure 4.4.

Problem 4.1. Produce the same figures for the MNIST dataset.

Problem 4.2. Does the result in Figure 4.4 show a good use of the latent space?

Chapter 5

Probabilistic Principal Components Analysis and Variational Autoencoder

5.1 Probabilistic PCA

In preparation for the idea of generative models, we would like to switch our point of view from representation of the data to representation of its distribution. This Bayesian idea, though not new in ML, turned out to be very powerful for NN as it leads to a whole new generation of machine learning algorithms that first used sampling methods to establish the learning processes and later used variational methods, where some of the units in the network became stochastic units that, strictly speaking, are non-differentiable. Thus, a statistical approach to NN comes from the network-fitting parameters of a probability distribution rather than from fitting the data. As a way to introduce such methods, we consider a Gaussian model for a probabilistic version of PCA where we assume that our low-dimensional feature vectors lie on a low-dimensional subspace where they are distributed as a zero mean and uncorrelated Gaussian (spherical distribution). The model assumes that the observed data were generated by a K-dimensional Gaussian variable $\nu \sim \mathcal{N}(\mathbf{0}, \mathbf{I}_K)$ that was then transformed up into N-dimensions, $\mathbf{x} = \mathbf{W}\nu + \mu$, where $\mathbf{W}$ is a $N \times K$ matrix. This transforms our initial zero mean and unit variance variable ν into a new variable $\mathbf{x} \sim \mathcal{N}(\mu, \mathbf{W}\mathbf{W}^\top)$. So, effectively, our low-dimensional randomness is "spread out" over many more dimensions (from K to N), but if we look at the covariance matrix of the data, it is low rank because it has only K independent rows or columns. We are in a very similar situation to deterministic

PCA where all vectors $\mathbf{x}$ generated from this model lie on a linear subspace of K dimensions.

Since real data of N dimensions rarely lie exactly on a lower-dimensional subspace, the full Gaussian model actually loads an extra diagonal matrix to the $\mathbf{WW}^\top$ matrix, $\mathbf{x} \sim \mathcal{N}(\boldsymbol{\mu}, \mathbf{WW}^\top + \sigma^2 \mathbf{I})$. In other words, we added uncorrelated noise to all dimensions to account for spread of the data over the whole N-dimensional space.

The importance of Probabilistic PCA (PPCA) is not so much in the model itself but in the conceptual shift of how we estimate the model. Instead of fitting the data, we actually want to fit a probability function that is a multi-variate Gaussian of K dimensions (how to know what is K is a separate question) so that the likelihood of our data in that probability will be maximized. This introduces a maximum-likelihood way of thinking about fitting generative models to data.

Given a dataset $X = \{\mathbf{x}_n\}$ of observed data points, its log likelihood function is given by

$$\ln p(X|\boldsymbol{\mu}, \mathbf{W}, \sigma^2) = \sum_{n=1}^{N} \ln p(\mathbf{x}_n|\boldsymbol{\mu}, \mathbf{W}, \sigma^2) \tag{5.1}$$

$$= \frac{NK}{2}\ln(2\pi) - \frac{N}{2}\ln|\mathbf{C}|$$

$$- \frac{1}{2}\sum_{n=1}^{N}(\mathbf{x}_n - \boldsymbol{\mu})^\top \mathbf{C}^{-1}(\mathbf{x}_n - \boldsymbol{\mu}), \tag{5.2}$$

where $\mathbf{C} = \mathbf{WW}^\top + \sigma^2 \mathbf{I}$. The features or latent variables in PPCA are the vectors $\boldsymbol{\nu}$ that are estimated using Bayes' rule

$$\boldsymbol{\nu} \sim p(\boldsymbol{\nu}|\mathbf{x}) = \mathcal{N}((\mathbf{WW}^\top + \sigma^2 \mathbf{I})^{-1}\mathbf{W}^\top(\mathbf{x} - \boldsymbol{\mu}),$$

$$(\mathbf{WW}^\top + \sigma^2 \mathbf{I})^{-1}\sigma^2). \tag{5.3}$$

It should be noted that for $\sigma^2 \to 0$, the mean parameter is a pseudo-inverse matrix multiplication of the data vector $\mathbf{x}$ with zero mean (i.e., with mean vector subtracted from the data). Moreover, Tipping and Bishop (1999) showed that the maximum of the likelihood function is obtained for a matrix $W_{ML} = \mathbf{U}_K(\boldsymbol{\Lambda}_K - \sigma^2 \mathbf{I})^{1/2}$, up to an arbitrary rotation (omitted here), with $\mathbf{U}_K$ being a matrix comprised of K eigenvectors with the largest K eigenvalues and $\boldsymbol{\Lambda}_K$ the diagonal matrix containing these eigenvalues.

Intuitively speaking, for $\sigma^2 \to 0$, the matrix that projects data into the latent space is exactly the same as in the deterministic PCA case, with the vectors scaled by a root of their respective eigenvalues. This scaling is needed since unlike the PCA case, where different dimensions appear with different scales, here we project everything into a spherical latent variable with unit variance, so each dimension needs to be re-scaled according to its variance.

5.2 Variational Autoencoder (VAE)

The case of PPCA in the previous section introduced the idea of maximum likelihood estimation, or in other words, optimizing the parameters of our learning model to find a probability distribution where the true data samples will be highly likely. This is a very different approach to learning compared to deterministic setting. Instead of applying some optimal transformation to the data itself, the goal of probabilistic modeling is to find an approximation to distribution of data in the real world. For example, in the case of a language model, maximum likelihood model will assign high probability to real sentences and low probability to badly structured or meaningless text. In fact, we want to be able to judge what events are probable and what are improbable in the world where the data originates from.

In this setting, encoding of data $\mathbf{x}$ into a latent representation $\mathbf{z}$ is done by posterior estimation $p(\mathbf{z}|\mathbf{x})$. The whole process of modeling becomes a process of probability approximation, done by searching for the best probability model by optimizing its parameters. In the case of variational AE (VAE), these parameters are learned by an NN.

So, how do we build a probabilistic model with a neural network? The idea is to use again an encoder and decoder network, but instead of using the encoder to find a latent state that is a lower-dimensional projection of the data, the encoder is used to find *estimates of the mean and variance of a multivariate Gaussian*, from which the latent variable is sampled. Then, this variable is passed to the decoder to reconstruct the input. For instance, if we consider a one-layer encoder and a one-layer decoder, an example of VAE is written as

$$p(\mathbf{x}|\mathbf{z}) = g(\mathbf{W}\mathbf{z} + \boldsymbol{\mu}), \tag{5.4}$$

$$q(\mathbf{z}|\mathbf{x}) = \mathcal{N}(g(\mathbf{V}_{\mu}\mathbf{x}), g(\mathbf{V}_{\Sigma}\mathbf{x})), \tag{5.5}$$

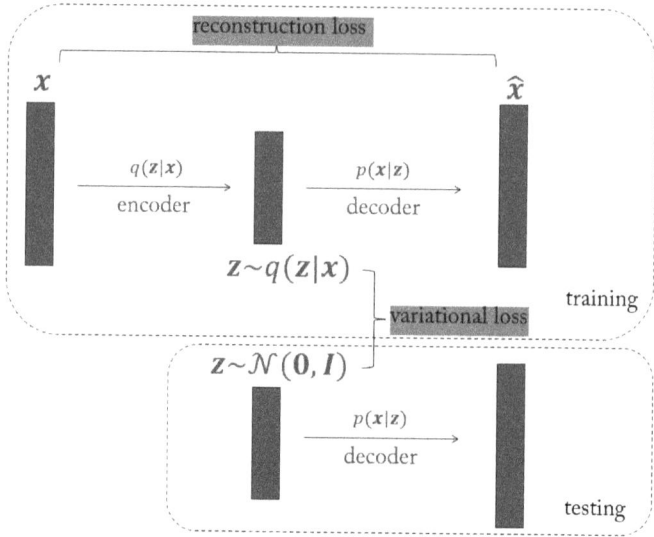

Fig. 5.1. The training and testing phases of VAE.

where without loss of generality we assume that $\mathbf{z}$ is zero mean (otherwise, the mean has to be subtracted from the input to the encoder q and added back in the decoder p). Also, we understand $p(\mathbf{x}|\mathbf{z})$ as a distribution with mean given by $g(\mathbf{W}\mathbf{z} + \boldsymbol{\mu})$ (the other information does not matter in VAE). One should note that the model that such an AE implements in the case of linear neurons (i.e., activation $g(a) = a$) is exactly PPCA.

Recall that AE is trained by minimizing the reconstruction loss. When training VAE, in addition to a similar reconstruction loss, we also need to minimize a variational loss which penalizes the latent distribution of $q(\mathbf{z}|\mathbf{x})$ from being away from a prior distribution, e.g., a standard normal distribution $\mathcal{N}(\mathbf{0}, \mathbf{I})$. We introduce the details of this variational loss in the following chapter. Once a VAE is trained, it can be used to sample from the data distribution that we have learned. This can be done by sampling from the prior distribution, again usually taken to be $\mathcal{N}(\mathbf{0}, \mathbf{I})$, and then passing it through the decoder. The training and testing phases of VAE are illustrated in Figure 5.1. Since the testing phase of VAE can be used as a generative model, we discuss the implementation of VAE later in Chapter 6, together with other generative models.

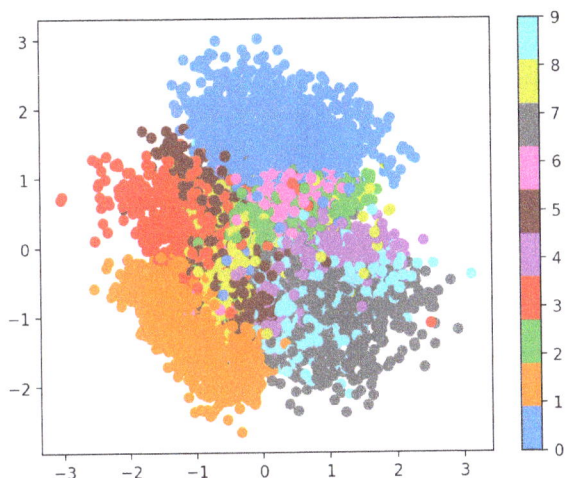

Fig. 5.2. Latent space of VAE for the MNIST data.

The training phase of VAE is very similar to an AE, except that we have a "variational loss" that compares $q(\mathbf{z}|\mathbf{x})$ with $\mathcal{N}(0, \mathbf{I})$. We expect that the latent space will be similar to a normal distribution if this loss is small. Figure 5.2 shows the latent space of a VAE for the MNIST data. Indeed, the distribution looks similar in all directions. An eager reader can refer to Chapter 6 for an implementation and then try to reproduce the same figure, or for other datasets.

Chapter 6

Variational Autoencoder

6.1 KL Divergence and Log Likelihood

To start the process of probability function approximation, we need to introduce some statistical notations:

- Kullback–Leibler (KL) divergence, which is a measure of discrepancy between two distributions:

$$\text{KL}(P, Q) = \sum p(x_i) \log \frac{p(x_i)}{q(x_i)}. \tag{6.1}$$

- Entropy H of a distribution:

$$H(\mathbf{x}) = -\sum_i p(x_i) \log p(x_i) \tag{6.2}$$

and its conditional version for the case of two variables:

$$H(\mathbf{x}|\mathbf{y}) = -\sum_{i,j} p(x_i, y_j) \log p(x_i|y_j). \tag{6.3}$$

- Mutual information between two random variables:

$$I(\mathbf{x}, \mathbf{y}) = H(\mathbf{x}) + H(\mathbf{y}) - H(\mathbf{x}, \mathbf{y})$$
$$= H(\mathbf{x}) - H(\mathbf{x}|\mathbf{y}) = H(\mathbf{y}) - H(\mathbf{y}|\mathbf{x}). \tag{6.4}$$

One can check that for $P = Q$, the value of their KL divergence is zero. KL is not formally a distance since it is not symmetric, but a symmetric version of it, known as Jensen–Shannon (JS) divergence

$$JS(P,Q) = \frac{KL(P,Q) + KL(Q,P)}{2}, \tag{6.5}$$

is possible too.

Few notes about the definitions and notations. The sum in the above expressions is over all possible realizations of the outcome of the random variables $\mathbf{x}$ and $\mathbf{y}$. This should not be confused with summing over samples or data points collected from a corpus. With sufficient samples, mean can be approximated by an average, but as far as the definitions themselves go, the elements in the summation are log probabilities, i.e., the weights in the summations $p(x_i)$ or $p(x_i, y_j)$ are probabilities and the sums are over the possible values of the variables x and y.

Another comment is about the ambiguity or duality in understanding these expressions as functions of probability or of the random variable. In many respects, it is really context dependent. Writing $H(\mathbf{x})$ or $H(P)$ where $\mathbf{x} \sim P(\mathbf{x})$ is equivalent. A confusion might occur when we think about encoder and decoder as $q(\mathbf{z}|\mathbf{x})$ and $p(\mathbf{x}|\mathbf{z})$, respectively. In fact, what we have are two different distributions, each one linking the latent variable $\mathbf{x}$ and $\mathbf{z}$ through a joint probability $P = p(\mathbf{x}, \mathbf{z})$ and $Q = q(\mathbf{x}, \mathbf{z})$. As we see in the following, Q is a *variational* approximation of P, where the neural network serves as a sort of so-called "Markov Blanket" that makes the transition from the outside world of true data samples to the internal world of the model defined by its latent states possible. This transition or mapping is done through function approximation, which we consider as an encoder $q(\mathbf{z}|\mathbf{x})$. Then, during the learning process, we decode the data and check it against the outside world statistics; we want to use in our calculations the true probabilities $p(\mathbf{x}|\mathbf{z})$.

Without further proof, we want to state some properties of these functions. To use the same notation as in the following VAE discussion, we use the pairs $(\mathbf{x}, \mathbf{z})$ instead of $(\mathbf{x}, \mathbf{y})$ that we used before:

$$KL(q(\mathbf{z}|\mathbf{x}), p(\mathbf{z}|\mathbf{x})) = KL(q(\mathbf{z}|\mathbf{x}), p(\mathbf{x}, \mathbf{z})) + \log p(\mathbf{x}), \tag{6.6}$$

$$KL(q(\mathbf{z}|\mathbf{x}), p(\mathbf{x}, \mathbf{z})) \overset{(1)}{=} KL(q(\mathbf{z}|\mathbf{x}), p(\mathbf{z})) - \mathbb{E}_{q(\mathbf{z}|\mathbf{x})}[\log p(\mathbf{x}|\mathbf{z})] \tag{6.7}$$

$$\overset{(2)}{=} \mathbb{E}_{q(\mathbf{z}|\mathbf{x})}[\log p(\mathbf{x}, \mathbf{z})] - H_{q(\mathbf{z}|\mathbf{x})}(q(\mathbf{z}|\mathbf{x})). \tag{6.8}$$

To derive all of these proofs, simply apply Bayes' rule. Let us now explain what these expressions mean. The first expression links the encoder to a true posterior of the data. In other words, we want to compare the distribution of the latent variables $\mathbf{z}$ in the encoder $q(\mathbf{z}|\mathbf{x})$ (i.e., when $\mathbf{x}$ is known), to its true conditional probability $p(\mathbf{z}, \mathbf{x})$. One should be careful here to realize that we have no way of experimentally observing the true $\mathbf{z}$, but at least mathematically, this is well defined and is used in subsequent derivations.

The first equality (1) of the second expression further allows us to relate the two quantities (encoding vs posterior) to a prior distribution of the latent variables $p(\mathbf{z})$. This prior $p(\mathbf{z})$ is where our assumptions about the nature of the world come in. We cannot observe $p(\mathbf{z}|\mathbf{x})$, but we can guess or assume that the latent or "hidden" causes in the real world that are responsible for generating the observations $\mathbf{x}$ behave in a certain way. This is a very strong assumption, and in most VAE applications, it is limited to a multi-variate, zero mean, and uncorrelated Gaussian distribution, which allows us to derive some powerful learning rules, as we soon see.

We also wrote a second equality (2) that relates encoding and the true world probability of the complete model (data, latent states) = $p(\mathbf{x}, \mathbf{z})$ to the expectation of this joint probability minus the entropy of the encoder $q(\mathbf{z}|\mathbf{x})$. The reason to use this second form is that it was common in early deep learning literature to assume that the joint probability of the (data, latent states) is governed by some *Energy* function, which then governs the probability as a negative exponent in a function:

$$p(\mathbf{x}, \mathbf{z}) = \frac{\exp(-\beta \cdot \mathrm{Energy}(\mathbf{x}, \mathbf{z}))}{Z}, \qquad (6.9)$$

where the capital Z is a normalization called a partition function which is basically a summation of the numerator over all external ($\mathbf{x}$) and internal ($\mathbf{z}$) system states. This way of writing probability makes expression (2) to be *Energy* $- \frac{1}{\beta}$*Entropy*, which is known in physics as *Free Energy*. So, there is a whole branch of deep learning that is inspired by statistical physics that talks about deep learning models as thermodynamic systems that live in balance with the environment, which is the data they see. Even more surprisingly, these models found their way into Neuroscience and led to emergence of a whole new field called "Free Energy of the Brain".

Summing these relations once more, we get a new expression for log-likelihood:

$$\log p(\mathbf{x}) = \mathrm{KL}(q(\mathbf{z}|\mathbf{x}), p(\mathbf{z}|\mathbf{x})) - \mathrm{KL}(q(\mathbf{z}|\mathbf{x}), p(\mathbf{z})) + \mathbb{E}_{q(\mathbf{z}|\mathbf{x})}[\log p(\mathbf{x}|\mathbf{z})] \tag{6.10}$$

$$= \mathrm{KL}(q(\mathbf{z}|\mathbf{x}), p(\mathbf{z}|\mathbf{x})) - \mathbb{E}_{q(\mathbf{z}|\mathbf{x})}[\log p(\mathbf{x}, \mathbf{z})] + H_{q(\mathbf{z}|\mathbf{x})}(q(\mathbf{z}|\mathbf{x})). \tag{6.11}$$

Since KL is greater or equal to zero, this equation is often rewritten as an upper bound, removing the $\mathrm{KL}(q(\mathbf{z}|\mathbf{x}), p(\mathbf{z}|\mathbf{x}))$ part:

$$\log p(\mathbf{x}) \geq \mathbb{E}_{q(\mathbf{z}|\mathbf{x})}[\log p(\mathbf{x}|\mathbf{z})] - \mathrm{KL}(q(\mathbf{z}|\mathbf{x}), p(\mathbf{z})). \tag{6.12}$$

The right-hand side of (6.12) is often called the Evidence Lower Bound (ELBO). The nice property of ELBO is that it can be computed using various approximation methods, and VAE is one of them (the other method called Expectation Maximization (EM) does it in iterative manner, but we won't cover it right now).

6.2   Computing the Evidence Lower Bound (ELBO)

What we have now is the following problem setting: Given some data distributed in the real world according to some unknown probability $p(\mathbf{x})$, we assume that there is a latent model that governs these data, expressed as $p(\mathbf{x}, \mathbf{z})$. In order to learn this model, we propose another more tractable distribution $q(\mathbf{x}, \mathbf{z})$ that can be effectively estimated using some optimization method, in our case, gradient descent implemented using a neural network. For the sake of completeness, let us write the optimization criteria here again:

$$\mathrm{ELBO} = \mathbb{E}_{q(\mathbf{z}|\mathbf{x})}[\log p(\mathbf{x}|\mathbf{z})] - \mathrm{KL}(q(\mathbf{z}|\mathbf{x}), p(\mathbf{z})). \tag{6.13}$$

Let's now spell out what we have here. Going back to the idea of VAE, $p(\mathbf{x}|\mathbf{z})$ is the decoder and $q(\mathbf{z}|\mathbf{x})$ is the encoder. Our neural networks will be trained so as to maximize ELBO, which amounts to two separate tasks:

(1) Maximize the likelihood of the data $\mathbf{x}$ that was obtained from latent states $\mathbf{z}$ encoded by our encoder network and

(2) Minimize the KL distance between the latent states that our encoder finds and the prior distribution of the latent state in the real world $p(\mathbf{z})$ (reminder: this prior distribution is our modeling assumption).

So, we have cross-entropy criteria applied to the output of VAE and KL criteria applied for estimating the stochastic part of the encoder. This way of deriving an objective learning function is known as *variational principle* of estimating generative models. Also, because ELBO can be written as free energy, it is also known as the *variational free energy*.

6.3 NN Implementation of VAE

To facilitate backpropagation, the implementation of VAE is done through so-called "reparametrization" trick, which led to the modern VAE popularity. Instead of estimating the encoder by sampling methods, the normal distribution of the latent prior $\mathbf{z}$ is distributed with a mean μ and variance σ^2. This is written as a transformation of a zero mean and unit variance Gaussian variable $\epsilon \sim \mathcal{N}(0, \sigma^2)$ as $x = \sigma^2 \epsilon + \mu$. This can be effectively implemented by drawing a random variable ϵ and then multiplying its value by variance and adding the mean. Figure 6.1 illustrates the above reparametrization process.

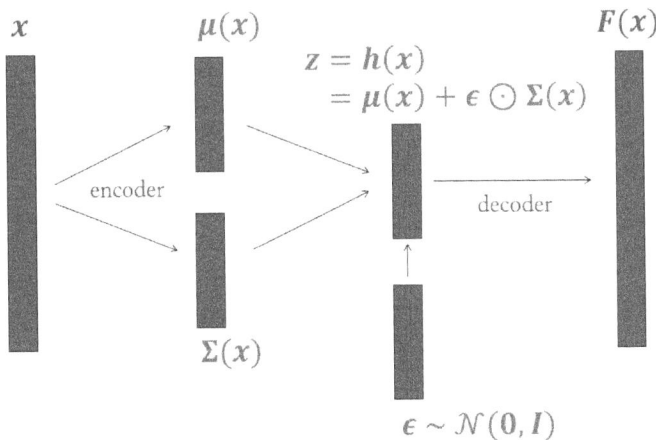

$$x \qquad \mu(x) \qquad F(x)$$

$$z = h(x) = \mu(x) + \epsilon \odot \Sigma(x)$$

encoder

decoder

$$\Sigma(x)$$

$$\epsilon \sim \mathcal{N}(0, I)$$

Fig. 6.1. The reparametrization trick of VAE.

This expression can be generalized for multi-variate Gaussians with a vector mean and covariance matrix. The parameters of the prior distribution are estimated by minimizing the KL distance between the encoding and prior. The KL objective can be explicitly written as KL between the zero mean and unit variance Gaussian that we assume to be the prior of $\mathbf{z}$ and the estimated posterior through the decoder $q(\mathbf{z}|\mathbf{x})$:

$$\mathrm{KL}(\mathcal{N}(\mu, \sigma) \parallel \mathcal{N}(0, 1)) = \sum_{\mathbf{x} \in X} \left(\sigma^2 + \mu^2 - \log \sigma - \frac{1}{2} \right), \qquad (6.14)$$

which is a particular case of a general KL distance between two Gaussian distributions:

$$\mathrm{KL}(p, q) = - \int p(x) \log q(x) dx + \int p(x) \log p(x) dx \qquad (6.15)$$

$$= \frac{1}{2} \log(2\pi\sigma_2^2) + \frac{\sigma_1^2 + (\mu_1 - \mu_2)^2}{2\sigma_2^2} - \frac{1}{2}(1 + \log 2\pi\sigma_1^2)$$

$$= \log \frac{\sigma_2}{\sigma_1} + \frac{\sigma_1^2 + (\mu_1 - \mu_2)^2}{2\sigma_2^2} - \frac{1}{2}.$$

The implementation of VAE uses two neural networks as function approximations for estimators of the mean and variance parameters, where the KL expression serves as the loss function for the back-propagation procedure. One note of caution needs to be said about what precedes these networks. The input data are encoded by a neural network that maps $\mathbf{x}$ to $\mathbf{z}$, and then the mean and variance are obtained by two linear fully connected networks that implement the reparametrization trick. The kind of encoding from $\mathbf{x}$ to $\mathbf{z}$ depends on the type of data, and it can be as simple as a single dense layer or have multiple layers in a complex deep convolutional network. Moreover, since $P(\mathbf{x}|\mathbf{z})$ at the output of the network is Gaussian, the cross-entropy at the output can be approximated by averaging the Euclidean distance between the decoded and true data values, an expression that appears in the exponent of the Gaussian probability function. Since $p(\mathbf{x}|\mathbf{z}) \sim \mathcal{N}(\mu, \sigma^2 \mathbf{I})$, we may estimate the mean parameter by averaging over the data reconstructed by the decoder $\hat{\mu} = \mathbb{E}_{q(\mathbf{z}|\mathbf{x})}[\hat{\mathbf{x}}]$. Accordingly, in order to estimate $\mathbb{E}_{q(\mathbf{z}|\mathbf{x})}[\log p(\mathbf{x}|\mathbf{z})]$, we take an approximate mean by averaging over multiple encoding–decoding reconstruction samples $\hat{\mathbf{x}}$:

$$\log p(\mathbf{x}|\mathbf{z}) \sim \log \exp(-\|\mathbf{x} - \hat{\boldsymbol{\mu}}\|^2) \sim -\sum(\mathbf{x} - \hat{\mathbf{x}})^2, \qquad (6.16)$$

where the sum is over a batch of data training examples $\mathbf{x}$ that are passed through VAE to reconstruct $\hat{\mathbf{x}}$. For more rigorous details about the specific implementation, see the work of (Kingma and Welling, 2014, Appendix C.2) in the original VAE paper where they talked about implementing Gaussian MLP as encoder–decoder.

The following is a sample code script for implementing a simple VAE using MLP in PyTorch:

```python
import torch
import torch.nn as nn

class VAE(nn.Module):
    def __init__(self, input_dim, hidden_dim, latent_dim):
        super(VAE, self).__init__()

        self.fc1 = nn.Linear(input_dim, hidden_dim)
        self.fc21 = nn.Linear(hidden_dim, latent_dim)
        self.fc22 = nn.Linear(hidden_dim, latent_dim)
        self.fc3 = nn.Linear(latent_dim, hidden_dim)
        self.fc4 = nn.Linear(hidden_dim, input_dim)

    def encode(self, x):
        h1 = nn.ReLU(self.fc1(x))
        return self.fc21(h1), self.fc22(h1)

    def reparameterize(self, mu, logvar):
        std = torch.exp(0.5*logvar)
        eps = torch.randn_like(std)
        return mu + eps*std

    def decode(self, z):
        h3 = nn.ReLU(self.fc3(z))
        return torch.sigmoid(self.fc4(h3))

    def forward(self, x):
        mu, logvar = self.encode(x.view(-1, 784))
        z = self.reparameterize(mu, logvar)
        return self.decode(z), mu, logvar
```

```
# let's consider the Fashion-MNIST data used in Chapter 1
input_dim = 784 # each image is 28x28
hidden_dim = 256
latent_dim = 16

model = VAE(input_dim, hidden_dim, latent_dim)

criterion = nn.BCELoss(reduction='sum')
optimizer = torch.optim.Adam(model.parameters(), lr=0.001)

epochs = 200

# assuming train_loader is a dataloader for the images
for epoch in range(epochs):
    model.train()
    train_loss = 0
    for batch_idx, (data, _) in enumerate(train_loader):
        data = data.view(-1, 784)
        optimizer.zero_grad()
        recon_batch, mu, logvar = model(data)
        loss = criterion(recon_batch, data) + 0.5 * torch.sum(1 +
            logvar - mu.pow(2) - logvar.exp())
        loss.backward()
        train_loss += loss.item()
        optimizer.step()
    print('Epoch {}: Average Loss: {:.4f}'.format(epoch, train_loss
        / len(train_loader.dataset)))
```

6.4   Estimating Model Parameters Using ELBO

The encoding of data in a unsupervised way is often done by using VAE. Alemi *et al.* (2018) showed that maximization of Evidence Lower Bound (ELBO) is equivalent to minimization of $I(X, Z)$ for the VAE encoder with an added minimization requirement of the decoder reconstruction error $D(X, Z)$. Let us consider the steps leading to this equivalence. We consider the encoding in terms of variational approximation $q(\mathbf{z}|\mathbf{x})$ for the encoder and true probability for the data $p(\mathbf{x})$. Accordingly, the mutual information between the data X and the latent states Z when using the encoder becomes a measure of for the joint distribution, expressed through the encoder $p_e(\mathbf{x}, \mathbf{z}) = p(\mathbf{x})q(\mathbf{z}|\mathbf{x})$, or through the decoder $p_d(\mathbf{x}, \mathbf{z}) = p(\mathbf{x}|\mathbf{z})q(\mathbf{z})$.

We assume that $p_e(\mathbf{x}, \mathbf{z}) = p_d(\mathbf{x}, \mathbf{z})$, so we use them interchangeably. The reason for explicitly writing the encoder–decoder joint distribution is to note the two ways of associating the data $\mathbf{x}$ with the latent representation $\mathbf{z}$ derived using the variational encoder–decoder approximation:

$$I(X, Z) = \int p_e(\mathbf{x}, \mathbf{z}) \log \frac{p_e(\mathbf{x}, \mathbf{z})}{q(\mathbf{z})p(\mathbf{x})} d\mathbf{z} d\mathbf{x} \qquad (6.17)$$

$$= \int p(\mathbf{x})q(\mathbf{z}|\mathbf{x}) \log \frac{q(\mathbf{z}|\mathbf{x})}{q(\mathbf{z})} d\mathbf{z} d\mathbf{x} = \mathbb{E}_{p(\mathbf{x})}[\mathrm{KL}(q(\mathbf{z}|\mathbf{x}), q(\mathbf{z}))]. \qquad (6.18)$$

Let us now introduce two auxiliary definitions: the reconstruction error

$$D = -\int p(\mathbf{x})q(\mathbf{z}|\mathbf{x}) \log p(\mathbf{x}|\mathbf{z}) = -\mathbb{E}_{p(\mathbf{x})}\mathbb{E}_{q(\mathbf{z}|\mathbf{x})}[\log p(\mathbf{x}|\mathbf{z})] \quad (6.19)$$

which is the average over all inputs $\mathbf{x} \sim p(\mathbf{x})$ of the negative log-likelihood of the reconstructed output using latent $\mathbf{z}$'s that were encoded by $q(\mathbf{z}|\mathbf{x})$ and a rate

$$R = \int p(\mathbf{x})q(\mathbf{z}|\mathbf{x}) \log \frac{q(\mathbf{z}|\mathbf{x})}{p(\mathbf{z})} = \mathbb{E}_{p(\mathbf{x})}[\mathrm{KL}(q(\mathbf{z}|\mathbf{x}), p(\mathbf{z}))] \qquad (6.20)$$

which is an average KL distance between the encoded $\mathbf{z}$'s and the prior $p(\mathbf{z})$. An important observation is that these two factors appear in the ELBO expression:

$$-\mathbb{E}_{p(\mathbf{x})}[\mathrm{ELBO}] = D + R. \qquad (6.21)$$

Using some algebraic manipulations we show in the appendix that

$$R \geq I(X, Z) \qquad (6.22)$$

or in summary $-\mathrm{ELBO} \leq I(X, Z) + D$.

By minimizing $-\mathrm{ELBO}$, we are minimizing an upper bound on the sum of encoding information $I_e(X, Z)$ and reconstruction error. Going back to the overall loss formulation in previous section, this combines the Information Bottleneck principle with an additional factor that captures the fidelity or quality of the representation $\mathbf{z}$ so that the latent representation does not distort the original signal

x too much. Neural network models that use this principle include β-VAE that adds a weighting factor β between R and D so that $\text{ELBO}(\beta) = D + \beta R$. A more general approach to using information interpretation of VAE is known as the InfoVAE family of neural networks.

If we review our specification of the goal of the learning system as one of finding an approximation to $P(Z|X)$ so that it is the most compact representation of X (i.e., $\min I(X, Z)$ that has a small representation error of X (i.e., $D(X, Z)$), while being the most information or revealing latent encoding of some other data Y (i.e., $\max I(Z, Y)$, we arrive at a combined expression

$$\min \mathcal{L} = \min(-\text{ELBO} + \gamma D_{\text{KL}}(p(Y|X)\|p(Y|Z))) \tag{6.23}$$

$$\geq \min(I(X, Z) + D(X, Z) - \gamma I(Z, Y)), \tag{6.24}$$

where we omitted once again the $I(X, Y)$ term that does not depend on the model parameters.

This gives us an insight into a combined principle for learning a representation Z that is formulated in terms of multiple simultaneous goals:

- finding the most efficient embedding or best compressed latent encoding (minimal $I(X, Z)$),
- this embedding best reconstructs the original data (minimal $D(X, Z)$), and
- finally, it also best predicts another data Y by having the smallest surprisal or most information about it (maximal $I(Z, Y)$).

PART 3

Convolutional, Recurrent, and Transformer Neural Networks

Chapter 7

Convolutional Neural Networks

7.1 Convolution

The advancements of deep learning have been largely driven by the widespread use of convolutional neural networks (CNNs). Convolution, as the name implies, plays a pivotal role in CNNs and are considered the fundamental operation in various architectures.

Convolution finds its most intuitive application in the domain of image processing. In this context, a grayscale image can be conceptualized as a function $f(x, y)$ over a two-dimensional space. At a specified location (x, y), the value of $f(x, y)$ represents the scalar pixel intensity. To emphasize the discrete nature of digital images, it is also convenient to use $I(i, j)$ to denote an image, where i is the index for rows of the image (the height dimension) and j is the index for the columns of the image (the width dimension). For instance, the left image in Figure 7.1 shows a grayscale photo of a kitchen. It is equivalent to the right image, which uses the vertical axis to denote the function value, over the two-dimensional domain.

Mathematically, the *convolution* of two functions f and h is defined to be

$$f * h(\mathbf{x}) = \int f(\mathbf{x} - \mathbf{u})h(\mathbf{u})d\mathbf{u} = \int f(\mathbf{u})h(\mathbf{x} - \mathbf{u})d\mathbf{u}. \qquad (7.1)$$

It is clear from the integral representation that in the convolution operation, f and h are symmetric. Nevertheless, in practical applications, it is often more convenient to treat them differently. In this

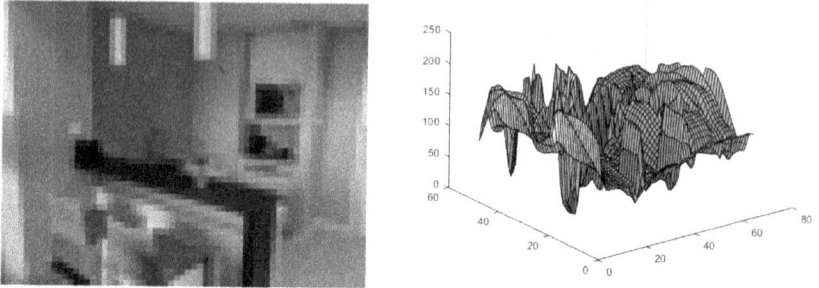

Fig. 7.1. An image as a function over a 2D space.

context, we commonly refer to f as the input and h as the (convolution) *kernel* or the (convolution) *filter*. This distinction helps clarify their respective roles in the convolution process.

In the case of discrete domains, such as when working with image data, we can replace the integral sign with the summation sign in the convolution operation without introducing any essential differences. For instance, let's consider an input 2D image I (represented as a function with a two-dimensional domain) and a 2D kernel K. The output feature map can be computed using the convolution operation as follows:

$$S(i,j) = (I * K)(i,j) = \sum_m \sum_n I(i-m, j-n)K(m,n). \qquad (7.2)$$

In the case of 1D signals, the discrete convolution can be implemented as a matrix multiplication operation. For instance, let's consider an input signal $\mathbf{x}$ of length n and a filter $\mathbf{h}$ of length m, where $m < n$. The convolution operation can be expressed as a matrix multiplication as follows:

$$\mathbf{x} * \mathbf{h} = \mathbf{A}\mathbf{x} = \begin{bmatrix} h_m & h_{m-1} & \cdots & h_1 & 0 & 0 & \cdots & 0 \\ 0 & h_m & h_{m-1} & \cdots & h_1 & 0 & \cdots & 0 \\ \vdots & \vdots & \ddots & & & & \ddots & \vdots \\ 0 & \cdots & 0 & h_m & h_{m-1} & \cdots & h_1 & 0 \\ 0 & \cdots & 0 & 0 & h_m & h_{m-1} & \cdots & h_1 \end{bmatrix} \begin{bmatrix} x_1 \\ x_2 \\ \vdots \\ x_n \end{bmatrix}.$$

$$(7.3)$$

Here, the matrix $\mathbf{A}$ is said to be a Toeplitz matrix because its entries satisfy $A_{i,j} = A_{i+1,j+1}$. In this case, the Toeplitz matrix is formed by shifting the filter $\mathbf{h}$ to different positions in each row while maintaining the same entries along each diagonal. The resulting matrix $\mathbf{A}$ captures the sliding behavior of the filter over the input signal. We remark that in 2D, it is still possible to represent convolution as a matrix multiplication using the so-called doubly block circulant matrices.

In a CNN, the kernel K is learned from data. Therefore, a specific ordering of the indices does not affect the result. Therefore, it is common to use a cross-correlation variant of the convolution operation, as opposed to a traditional convolution in (7.2). Specifically, the cross-correlation variant is implemented as:

$$S(i,j) = \sum_m \sum_n I(i+m-1, j+n-1)K(m,n). \qquad (7.4)$$

The specific limits of the summation indices m and n depend on the size and shape of the input feature map and the kernel, which is clear from the definition of the operation. For instance, let's consider the following scenario where the input feature map I has dimensions $N \times M$, where N represents the height and M represents the width of the input feature, and the kernel K has dimensions $k \times l$, where k represents the height and l represents the width of the kernel. In the cross-correlation operation, the output feature map S, computed by (7.4), has the following limits of indices:

- The limits of the summation indices m and n range from 1 to k and 1 to l, respectively. These limits ensure that the kernel K is properly aligned with the corresponding elements of the input feature map I during the cross-correlation operation.
- The range of i is from 1 to $N - k + 1$, whereas the range of j is from 1 to $M - l + 1$. This ensures that the kernel can be properly positioned within the height and width dimensions of the input feature map without exceeding its boundaries.

It's important to note that the actual limits may vary depending on the specific implementation or framework used, as some frameworks may adopt different conventions or indexing schemes.

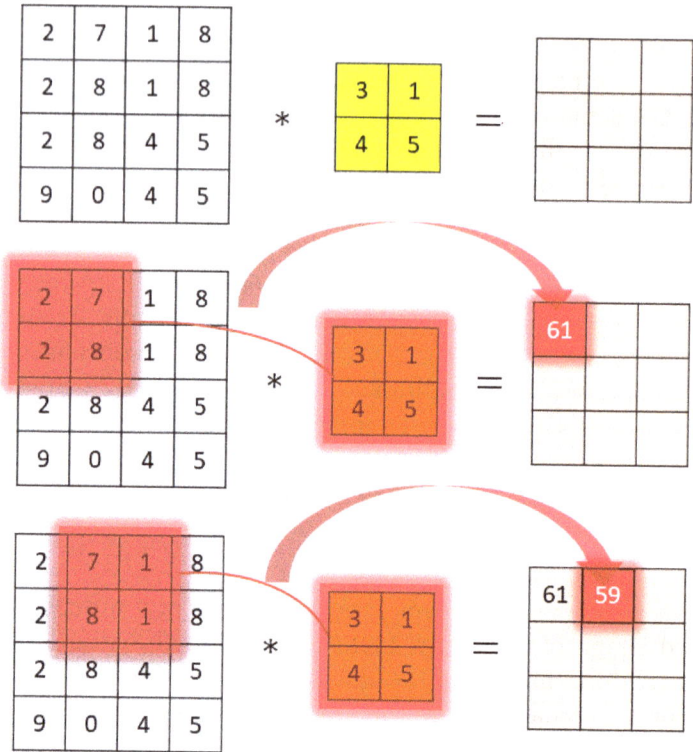

Fig. 7.2. An example illustrating the convolution operation.

Let's look at an example illustrated in Figure 7.2. In this example, the light blue 4×4 matrix is the input feature and the yellow 2×2 matrix is the kernel. Therefore, the output feature has dimension $(4 - 2 + 1) \times (4 - 2 + 1)$, which is the orange matrix shown on the right. To determine the entries of the output feature, we slide the kernel over the input feature. In this example, the first entry of the output feature matrix corresponds to aligning the kernel with the 2 submatrix in the top left corner of the input feature (the red shaded part in the second row of Figure 7.2). We take the sum of the pointwise products over all the locations from these two 2×2 matrix. This corresponds to the double summation $\sum_m \sum_n$ in (7.4), and in this example, it is $2 \cdot 3 + 7 \cdot 1 + 2 \cdot 4 + 8 \cdot 5 = 61$. After this is done, we move the kernel so that it is aligned with the red-shaded part in the third row of Figure 7.2 and we compute $7 \cdot 3 + 1 \cdot 1 + 8 \cdot 4 + 1 \cdot 5 = 59$ as the second entry of the output feature. By sliding the kernel over the entire input feature matrix, we can compute all the entries of the output feature matrix using this approach.

Problem 7.1. Complete the computation in Figure 7.2.

Problem 7.2. For discrete features in 1D, the convolution operation is defined as

$$y[n] \equiv (x * h)[n] = \sum_k x[k]h[n-k]. \tag{7.5}$$

(1) Compute convolution of the following features:
   (a) $x = [1, 2, 4]$, $h = [1, 1, 1, 1, 1]$.
   (b) $x = [1, 2, -1]$, $h = x$.
(2) If $y[n] = (x * h)[n]$, $n \in \mathbb{Z}$, show that

$$\sum_{n=-\infty}^{\infty} y[n] = \left(\sum_{n=-\infty}^{\infty} x[n] \right) \left(\sum_{n=-\infty}^{\infty} h[n] \right). \tag{7.6}$$

7.2 Motivation for Using Convolution in Neural Networks

Convolution leverages three important ideas in machine learning: sparse interactions, parameter sharing, and equivariant representations.

In contrast to fully connected networks where every neuron is connected to all neurons in the previous and next layers, CNNs exploit *sparse* interactions. This allows them to capture local features in the input, such as specific parts of an image. By using convolution kernels that are smaller than the input, CNNs focus on local receptive fields, making them well suited for tasks that emphasize local patterns.

Parameter sharing is another key concept in CNNs. Since the convolution kernel is applied across the entire input, the same set of parameters is shared and used multiple times for different regions of the input. This parameter sharing leads to a strong regularization

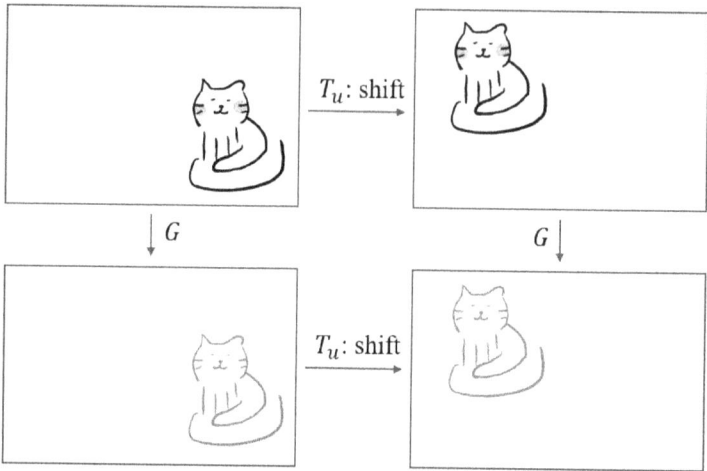

Fig. 7.3. Translation equivariance. Here, the function G is translation equivariant because it commutes with the shift. That is, the shift followed by G is equivalent to G followed by the shift. The image used in this example is credited to Yijia Xue.

effect, as well as computational efficiency. By reusing the parameters, CNNs reduce the number of learnable parameters, enabling more efficient training and inference.

Equivariant representation is particularly intuitive for image data. It means that if the input of a CNN is translated or shifted, the output will translate or shift in the same way. Figure 7.3 is a simple example that illustrates *translation equivariance*. In the top row of Figure 7.3, a cat is shifted from the bottom right corner to the top left corner of the input image. In the bottom row, the output of a function G also shifts in the same manner. The function G is said to be translation equivariant if the same holds for any input image and any shift.

Mathematically, let $\mathcal{T}_{\mathbf{u}}$ denote translation by $\mathbf{u}$ so that $\mathcal{T}_{\mathbf{u}}f(\mathbf{x}) := f(\mathbf{x} - \mathbf{u})$. For convolution, we have for any f and any $\mathbf{u}$ that

$$(\mathcal{T}_{\mathbf{u}}f) * h = \mathcal{T}_{\mathbf{u}}(f * h). \tag{7.7}$$

Therefore, convolution is translation equivariant.

Problem 7.3. Prove (7.7).

The idea of convolution is not only used in neural networks but widely applied in other areas. A typical example is a wavelet transform. In particular, Mallat (2012) defined a multi-layer and multi-scale feature extractor, called the scattering transform, the filters of which are constructed using wavelets. Since its structure is very similar to a neural network except for that the weighted are fixed in the filters, the scattering transform is often used to mathematically interpret and analyze properties of CNNs.

7.3 Multichannel Convolution

In many applications, the input and the output features of a convolution layer both have multiple channels. A common example of a multi-channel input is a color image, which typically consists of three channels: red, green, and blue (RGB). Each channel represents the intensity or color information for a specific color component. Similarly, the output feature of the convolution layer may also have multiple channels. Each channel in the output represents a different feature or filter learned by the CNN. These channels capture different aspects of the input, such as edges, textures, or higher-level semantic features.

If both input and output features are 3D tensors (two dimensions for height and width and the remaining dimension for channels), the convolution kernel has to be a 4D tensor $\mathbf{K}$. Denote the (i, j, k, l)th entry of $\mathbf{K}$ by $\mathbf{K}_{i,j,k,l}$, where i is the index for the channel of the output, j is the index for the channel of the input, and k and l are the indices in a regular 2D convolution kernel.

Suppose the input is a 3D tensor $\mathbf{V}$ with entries $\mathbf{V}_{j,k,l}$, where j represents the index of the channel of the input. Also, suppose the output is a 3D tensor $\mathbf{Z}$ with entries $\mathbf{Z}_{i,k,l}$, where i represents the index of the channel of the output. Then, the convolution $\mathbf{Z} = \mathbf{V} * \mathbf{K}$ is implemented as follows:

$$\mathbf{Z}_{i,k,l} = \sum_{j,m,n} \mathbf{V}_{j,k+m-1,l+n-1} \mathbf{K}_{i,j,m,n}. \qquad (7.8)$$

In practice, it is sometimes useful to consider a bias term for each output channel, similarly to fully connected networks. That is, we can use the affine transform $\mathbf{Z} = \mathbf{V} * \mathbf{K} + \mathbf{b}$, which reads

$$\mathbf{Z}_{i,k,l} = \left(\sum_{j,m,n} \mathbf{V}_{j,k+m-1,l+n-1} \mathbf{K}_{i,j,m,n} \right) + \mathbf{b}_i. \qquad (7.9)$$

In this case, we need to learn the entries of both $\mathbf{K}$ and $\mathbf{b}$. It is important to note that $\mathbf{b}_i$ is shared over all spatial locations (for all k and l).

7.4   Other Components of a Convolution Layer in a CNN

In PyTorch, a convolution layer can be defined using the following class (let's consider 2D convolution since this is the typical convolution used for image data):

```
CLASS torch.nn.Conv2d(in_channels, out_channels, kernel_size,
    stride=1, padding=0, dilation=1, groups=1, bias=True,
    padding_mode='zeros', device=None, dtype=None)
```

Here, `in_channels` is the number of channels in the input feature and `out_channels` is the number of channels in the output feature. They correspond to the total number of j and i indices, respectively, in the context of (7.8) and (7.9). `kernel_size` is the dimension of the convolution kernel. Next, we explain briefly the meaning of `stride` and `padding`.

The `stride` parameter determines the step size at which the convolution kernel slides over the input feature. In other words, it defines the spacing between consecutive kernel placements. `stride = 1` means that the kernel moves one unit at a time, resulting in overlapping receptive fields. Larger stride values, such as 2 or more, lead to a downsampling effect as the receptive fields become more sparse. The choice of stride affects the spatial resolution of the output feature map, with a larger stride resulting in reduced spatial dimensions. Figure 7.4 shows an example with `stride=2`. When the kernel slides over the input feature, it skips one pixel each time it is moved. Since the size of the input feature is 7×7 and the kernel size is 3×3, the output feature will have a size of $((7-3)/2+1) \times ((7-3)/2+1) = 3 \times 3$.

The `padding` operation refers to the technique of adding additional border pixels to the input feature. It is used to control the spatial

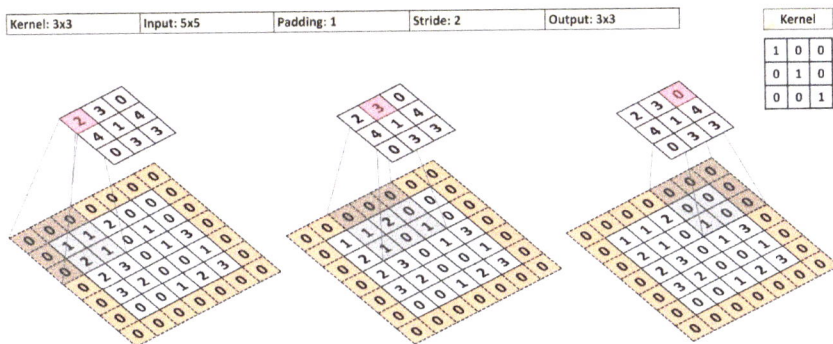

Fig. 7.4. An example of convolution with `kernel_size = 3`, `stride = 2`, and `padding = 1`. Each of the three subfigures shades a receptive field in the input feature and the corresponding pixel in the output feature.

dimensions of the output feature and prevent information loss at the borders. For instance, `padding=0` means no padding is applied, and `padding=1` means that a padding of one pixel is added to all sides of the input feature before performing convolution. The case `padding=1` is illustrated by the colored zeros in the input feature in Figure 7.4.

Typically, a convolution layer in a CNN consists of convolution kernels and possibly biases. Following the convolution, a nonlinear activation function is commonly applied. Additionally, a pooling operation, which can be treated as a separate layer, is often employed. The purpose of pooling is to reduce the dimensionality of its input by extracting one representative value from each local region of the input. One commonly used pooling operation is *max-pooling*, where the maximum value in each local region is extracted. Another possible choice is *average-pooling*, where the mean value is extracted from each region.

The pooling operation in CNNs serves multiple purposes. One of its benefits is reducing the complexity of the representation, making it more computationally efficient. However, another crucial aspect of pooling is its ability to introduce approximate translation invariance to the input, on top of the translation equivariance introduced by the convolution operation. This means that small translations in the input data have minimal impact on the pooled representation as long as the representative values remain within their respective local regions. Translation invariance is especially advantageous in tasks

```
================================================================
Layer (type)        Output Shape        Param #    Trainable
================================================================
Conv2d              64 x 4 x 14 x 14    40         True
```
```
ReLU                64 x 4 x 14 x 14    0          False
```
```
Conv2d              64 x 8 x 7 x 7      296        True
```
```
ReLU                64 x 8 x 7 x 7      0          False
```
```
Conv2d              64 x 16 x 4 x 4     1,168      True
```
```
ReLU                64 x 16 x 4 x 4     0          False
```
```
Conv2d              64 x 32 x 2 x 2     4,640      True
```
```
ReLU                64 x 32 x 2 x 2     0          False
```
```
Conv2d              64 x 2 x 1 x 1      578        True
```
```
Flatten             64 x 2              0          False
```

Fig. 7.5.   An output from a CNN for image classification.

like image classification. For example, if an image of a cat is translated to different locations within the image, a well-designed classifier should still recognize it as a cat and provide consistent predictions regardless of its position. By incorporating pooling operations, CNNs can achieve a certain degree of translation invariance, enabling them to handle variations in object position and maintain consistent classification performance.

Problem 7.4. Consider Figure 7.5. Suppose the input image is of size 28×28. Infer what kernels and operations are applied in each Conv2d layer.

7.5   Examples of CNNs

AlexNet: AlexNet is a CNN architecture that advanced the current development of deep neural networks. It was developed by Kirzhevsky *et al.* (2012) and won the first place in the ImageNet Large Scale Visual Recognition Challenge (ILSVRC) in 2012, with a large margin over the second-place model. The name "AlexNet" is after the first name of the first author of the paper.

Zeiler and Fergus (2014) analyzed a CNN very similar to AlexNet by applying the transpose of convolution operations and obtained visualizations of features in each layer. The conclusion is that shallower layers in a CNN tend to capture local and basic features, such as edges and textures; on the other hand, deeper layers tend to capture global semantics. This is not surprising since each convolution operation by definition is local, and only by cascading several layers of convolution can a neural network combine information from different regions of an image.

The original implementation of AlexNet requires distributed computation. However, looking back at AlexNet nowadays, it is rather simple to implement. Its pretrained version can also be easily loaded using PyTorch as follows:

```
import torch
model = torch.hub.load("pytorch/vision:v0.10.0", "alexnet",
    pretrained=True)
model.eval()
```

Note that `model.eval()` is needed since it claims that the model is used for evaluation instead of training. Recall that layers such as dropout and batch normalization behave differently in training and testing.

We can also implement AlexNet from scratch, which is not too much a headache and can serve as a good exercise. Note that on most machines we can implement a non-parallelized version directly. An example is as follows:

```
import torch
import torch.nn as nn

class AlexNet(nn.Module):
    def __init__(self, num_classes=1000):
        super(AlexNet, self).__init__()

        self.convs = nn.Sequential(
            nn.Conv2d(3, 96, kernel_size=11, stride=4, padding=2),
            nn.ReLU(inplace=True),
            nn.MaxPool2d(kernel_size=3, stride=2),

            nn.Conv2d(96, 256, kernel_size=5, padding=2),
            nn.ReLU(inplace=True),
```

```
            nn.MaxPool2d(kernel_size=3, stride=2),

            nn.Conv2d(256, 384, kernel_size=3, padding=1),
            nn.ReLU(inplace=True),

            nn.Conv2d(384, 256, kernel_size=3, padding=1),
            nn.ReLU(inplace=True),

            nn.Conv2d(256, 256, kernel_size=3, padding=1),
            nn.ReLU(inplace=True),
            nn.MaxPool2d(kernel_size=3, stride=2)
        )

        self.fcs = nn.Sequential(
            nn.Dropout(),
            nn.Linear(9216, 4096),
            nn.ReLU(inplace=True),

            nn.Dropout(),
            nn.Linear(4096, 4096),
            nn.ReLU(inplace=True),

            nn.Linear(4096, num_classes)
        )

    def forward(self, x):
        x = self.convs(x)
        x = torch.flatten(x, 1)
        x = self.fcs(x)

        return x
```

VGG: VGG is a widely used CNN, which was proposed by Simonyan and Zisserman (2015) and shorted came after AlexNet. VGG is named after the Visual Geometry Group at the University of Oxford. It also played a significant role in the development of CNNs, largely because it demonstrated the benefits of increased model depth. In the original paper, VGGs of different numbers (11, 13, 16, 19) of layers are considered. These models are nowadays widely known as VGG11, VGG13, VGG16, and VGG19. It is still simple to implement VGGs in PyTorch. For instance, we can directly load pretrained VGGs as follows:

```
import torch
model = torch.hub.load("pytorch/vision:v0.10.0", "vgg11",
    pretrained=True)
# or any of these variants
# model = torch.hub.load("pytorch/vision:v0.10.0", "vgg11_bn",
    pretrained=True)
# model = torch.hub.load("pytorch/vision:v0.10.0", "vgg13",
    pretrained=True)
# model = torch.hub.load("pytorch/vision:v0.10.0", "vgg13_bn",
    pretrained=True)
# model = torch.hub.load("pytorch/vision:v0.10.0", "vgg16",
    pretrained=True)
# model = torch.hub.load("pytorch/vision:v0.10.0", "vgg16_bn",
    pretrained=True)
# model = torch.hub.load("pytorch/vision:v0.10.0", "vgg19",
    pretrained=True)
# model = torch.hub.load("pytorch/vision:v0.10.0", "vgg19_bn",
    pretrained=True)
model.eval()
```

The following is an example of PyTorch implementation of VGGs from scratch:

```
import torch
import torch.nn as nn

class VGG(nn.Module):
    def __init__(self, num_classes=1000, cfg=None):
        super(VGG, self).__init__()

        self.convs = self._make_layers(cfg)
        self.fcs = nn.Sequential(
            nn.Linear(512 * 7 * 7, 4096),
            nn.ReLU(True),
            nn.Dropout(),
            nn.Linear(4096, 4096),
            nn.ReLU(True),
            nn.Dropout(),
            nn.Linear(4096, num_classes),
        )

    def forward(self, x):
        x = self.convs(x)
        x = torch.flatten(x, 1)
        x = self.fcs(x)
        return x
```

```
    def _make_layers(self, cfg):
        layers = []
        in_channels = 3

        for v in cfg:
            if v == 'M':
                layers.append(nn.MaxPool2d(kernel_size=2, stride=2))
            else:
                layers.append(nn.Conv2d(in_channels, v,
                    kernel_size=3, padding=1))
                layers.append(nn.ReLU(True))
                in_channels = v

        return nn.Sequential(*layers)


# VGG configurations
cfgs = {
    "VGG11": [64, 'M', 128, 'M', 256, 256, 'M', 512, 512, 'M', 512,
        512, 'M'],
    "VGG13": [64, 64, 'M', 128, 128, 'M', 256, 256, 'M', 512, 512,
        'M', 512, 512, 'M'],
    "VGG16": [64, 64, 'M', 128, 128, 'M', 256, 256, 256, 'M', 512,
        512, 512, 'M', 512, 512, 512, 'M'],
    "VGG19": [64, 64, 'M', 128, 128, 'M', 256, 256, 256, 256, 'M',
        512, 512, 512, 512, 'M', 512, 512, 512, 512, 'M'],
}

# Create VGG model
model = VGG(num_classes=1000, cfg=cfgs["VGG16"])
```

ResNet: ResNet is a deep CNN introduced by Ho *et al* (2016). The name "ResNet" is short for Residual Neural Network. The innovation over previous works is in its use of residual blocks and skip connections, which allows for the construction of very deep networks while mitigating the vanishing gradient problem. ResNet significantly advanced the field of computer vision and is nowadays still a very widely used and adapted model.

Figure 7.6 illustrates the key building block of a ResNet. If the input of the block is $\mathbf{x}$, instead of learning the desired output $\mathcal{H}(\mathbf{x})$, the weight layers will learn the residual function $\mathcal{F}(\mathbf{x}) = \mathcal{H}(\mathbf{x}) - \mathbf{x}$. The input $\mathbf{x}$ will have a skip connection with the residual so that they together add up to $\mathcal{F}(\mathbf{x}) + \mathbf{x}$.

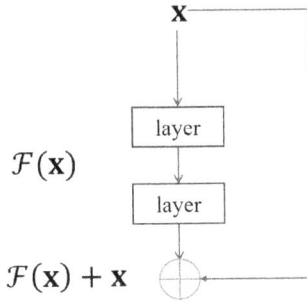

Fig. 7.6. Illustration of a building block of ResNet.

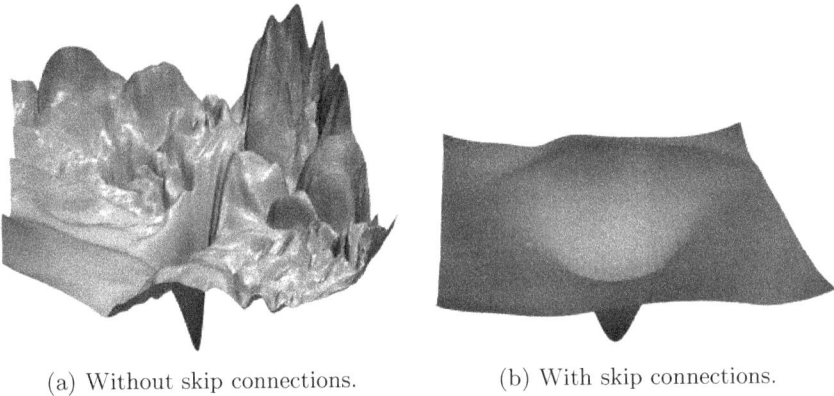

(a) Without skip connections.　　(b) With skip connections.

Fig. 7.7. Loss landscapes of a ResNet (Li *et al.*, 2018).

ResNet can contain a custom number of these building blocks. PyTorch contains the following widely adapted ResNet architectures:

```
import torch
model = torch.hub.load('pytorch/vision:v0.10.0', 'resnet18',
    pretrained=True)
# or any of these variants
# model = torch.hub.load('pytorch/vision:v0.10.0', 'resnet34',
    pretrained=True)
# model = torch.hub.load('pytorch/vision:v0.10.0', 'resnet50',
    pretrained=True)
# model = torch.hub.load('pytorch/vision:v0.10.0', 'resnet101',
    pretrained=True)
# model = torch.hub.load('pytorch/vision:v0.10.0', 'resnet152',
    pretrained=True)
model.eval()
```

Compared with regular CNNs, the ResNet skip connection helps obtain a smooth loss landscape. Li *et al.* (2018) introduced a tool for visualizing CNNs https://github.com/tomgoldstein/loss-landsca pe, which produced Figure 7.7, showing the loss landscapes without and with skip connections, respectively. It is clear that with skip connections as in ResNet, the loss landscape is much smoother and thus the corresponding network is easier to train. An interactive visualizer can be found at http://www.telesens.co/loss-landscape-viz/viewer. html.

Chapter 8

CNN Applications in Vision and Audio

8.1 CNN in Classification and Regression

One of the simplest use scenarios for CNNs is in the application of classification and regression problems. In this case, the convolution layers are typically used as feature extractors, which are followed by fully connected layers acting as classifiers or regressors. CNNs can achieve competitive performance on a wide range of classification and regression tasks. We briefly review the following applications.

Scene recognition: Scene recognition is the process of automatically identifying the type or category of a scene in an image or video. This involves analyzing the visual content of the scene, such as the presence of objects, textures, and patterns, as well as the spatial relationships and arrangements between them. Figure 8.1 is an example of input and possible outcome from a CNN for this task.

There are many public datasets available for training CNNs for scene recognition. For instance, Places365 (Zhou *et al.*, 2017) is a large-scale dataset which contains about 1.8 million images from 365 scene categories. Of course, as you can imagine, this dataset can be called from PyTorch using `torchvision.datasets.Places365`. There are also pretrained CNNs such as AlexNet and ResNet available at https://github.com/CSAILVision/places365.

Photo evaluation: Photo evaluation is another typical task that can be easily completed using CNNs. Unlike scene recognition which

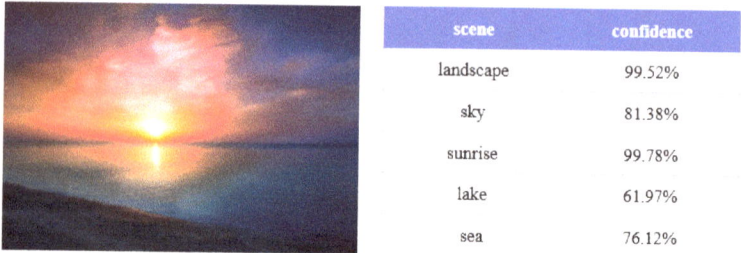

scene	confidence
landscape	99.52%
sky	81.38%
sunrise	99.78%
lake	61.97%
sea	76.12%

Fig. 8.1. A scene picture and possible outcome from CNN.

Fig. 8.2. Sample images from the CFP dataset. Each pair shows the frontal and profile images of the same person.

is a classification task, photo evaluation is a regression task which involves training a model on a large dataset of images labeled with quality scores and using the model to predict the quality of new images.

A typical dataset for photo evaluation is the Aesthetic Visual Analysis (AVA) dataset (Murray *et al.*, 2012), which contains more than 250,000 images, all annotated by a set of photographers. The annotation is on a scale of 1 to 10, which is used as the target for the regression task.

Face recognition: Face recognition involves identifying or verifying the identity of a person from his or her facial features. It has many applications, including ID verification, surveillance, and access control. A typical CNN-based approach is to capture an image of a person's face and use CNN to output the identity of the person, or to compare extracted features with features of known faces in a database.

Of course, we expect face recognition to be a more challenging task than scene recognition. This is because images of the same person may differ in pose, occlusion, or facial expression. For instance, Figure 8.2 shows several pairs of example from the Celebrities in

Frontal-Profile (CFP) dataset (Sengupta *et al.*, 2016), which contains images of human faces with both frontal and profile views. Given that frontal faces are more widely available in datasets, features from CNN may be biased toward them. In this case, it would be beneficial to learn to project the profile face into a frontal one. An example of such work is Cao *et al.* (2018), where a "head rotation estimator" is learned as residual so that final representations are all frontal. This greatly helps in the recognition task.

It is also worth mentioning that there are various attacking methods to face recognition systems. For instance, the face spoofing attacks try to trick the systems using fake images. Instead of images of real human faces, typical fake images include images of printed photos, images of phone or tablet screens showing photos, and images of masked human. Of course, such attacks will produce noises to face images and are detectable. There are many works on de-spoofing. An example is the work of Jourabloo *et al.* (2018), where a CNN model is trained to decompose a face image into a spoof noise and a human face.

There are many attacking methods to CNNs and spoofing attacks are just one of them. Therefore, it is very important to have robust CNNs. In an early paper, Szegedy *et al.* (2013) found that CNNs can be completely fooled by only adding a small noise. Later, Goodfellow *et al.* (2015) proposed a very simple way of fooling a neural network by injecting a noise. This is done by adding

$$\eta = \epsilon \operatorname{sign}(\nabla_{\mathbf{x}} \ell(\mathbf{x}, y; \boldsymbol{\theta})) \tag{8.1}$$

to the image $\mathbf{x}$. This is the earliest adversarial attack method to CNNs. In their example, a small noise added to a picture of a panda will be recognized as a gibbon by GoogleNet, with 99.3% confidence. Nowadays, various adversarial attack and defense methods have been proposed and it is still an important research field.

8.2 CNN in Object Detection

Object detection refers to detecting and localizing objects of interest within digital images and videos. The goal is to classify the objects as well as to provide precise bounding box coordinates around them. Figure 8.3 illustrates two typical outputs from objection detection.

Fig. 8.3. Two examples of object detection using YOLO. Top row: Original images; bottom row: Detection results.

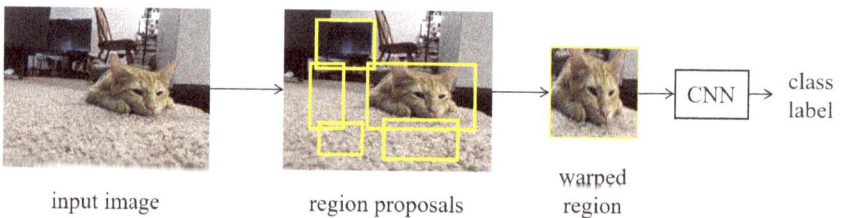

input image   region proposals   warped region   CNN → class label

Fig. 8.4. Illustration of R-CNN.

An early CNN-based approach to object detection is the Region-based CNN, or R-CNN for short, introduced by Girshick *et al.* (2015). R-CNN advanced the development of object detection by achieving significant improvement over earlier methodologies. It contains three steps: first, generating region proposals; second, extracting features; third, classifying regions and regressing bounding boxes. Figure 8.4 illustrates the main steps. In the first step, the regions are proposed via "selective search", which is a fixed algorithm with no training involved. Specifically, we first generate many candidate regions and then combine similar regions using a greedy algorithm. In the second step, each region is warped to a fixed size and a pretrained CNN is used to extract features. In the third step, classifier and regressors with fully connected layers are trained.

One problem with the original version of R-CNN is that it is relatively slow in training, due to the large number of proposed regions. Also, it takes tens of seconds for each test image at that time so that it cannot be implemented in real time. To this end, Girshick (2015) introduced an improved version called Fast R-CNN. Fast R-CNN uses shared convolutional layers to extract features only once for the entire image. Also, instead of warping the regions to a fixed size, Fast R-CNN divides each proposal into subregions of equal sizes and applies max pooling within each subregion to generate fixed-sized feature maps for each proposal. Therefore, Fast R-CNN is much more efficient than the original R-CNN. Moreover, Fast R-CNN enables end-to-end training and employs a loss for both classification and bounding box regression, which leads to an improvement of accuracies in detection and localization.

A later improvement to Fast R-CNN is the Faster R-CNN model, introduced by Ren *et al.* (2015). Unlike the original R-CNN or Fast R-CNN, Faster R-CNN does not apply the selective search method. Instead, it employs a network for the region proposal, which is considered as part of the overall network architecture. Faster R-CNN is widely adapted and we can use pretrained models from PyTorch. See http://pytorch.org/vision/master/models/faster_rcnn.html. An example is given as follows:

```python
from torchvision.models.detection import fasterrcnn_resnet50_fpn,
    FasterRCNN_ResNet50_FPN_Weights

weights = FasterRCNN_ResNet50_FPN_Weights.DEFAULT
transforms = weights.transforms()

images = [transforms(im) for im in images] # preprocess the input
    images

model = fasterrcnn_resnet50_fpn(weights=weights, progress=False)
model = model.eval()

outputs = model(images) # coordinates of bounding boxes together
    with class labels
```

Another popular object detection algorithm is the YOLO model. YOLO is short for "You Look Only Once", which highlights the fact that it uses a single neural network to predict bounding boxes and class labels directly from entire images in a single evaluation. It operates as follows. First, the input image is divided into $S \times S$ grids. For each grid, the neural network predicts the B bounding boxes (each box is determined by five numbers: the horizontal and vertical coordinates of an anchor point, the width and the height of the box, and a confidence score) and class probabilities for C classes. Therefore, we need a tensor of shape $S \times S \times B(5 + C)$ for these predictions.

Problem 8.1. The original YOLO model outputs a $7 \times 7 \times 30$ tensor for each input image. Suppose each cell corresponds to 2 box proposals. How many classes of objects are there in the dataset?

YOLO has many versions. All the versions follow the same strategy but differ in the backbone network, etc. In addition to the original YOLOv1, well-known models include YOLOv2 (Redmon and Farhadi, 2016) and YOLOv3 (Redmon and Farhadi, 2018). The history of YOLOv1–v8 is described at https://docs.ultralytics.com/, where YOLOv8 is introduced.

8.3   CNN in Segmentation

Image segmentation refers to dividing a digital image into multiple segments or regions, each consisting of a set of pixels. The purpose of segmentation is to simplify or transform the image representation into a more simple and meaningful form so that users can use it to locate objects, boundaries, and other important features within the image. Usually, image segmentation requires assigning a label to every pixel of an image and is thus a pixel-level task. It is also called semantic segmentation since the pixel classification reveals the semantics.

Traditionally, superpixel segmentation is widely used as preprocessing of image segmentation. It groups pixels with similar characteristics into compact regions called superpixels. For instance, Figure 8.5 shows examples of superpixel segmentation for the images in the top row of Figure 8.3. An example of segmentation methods

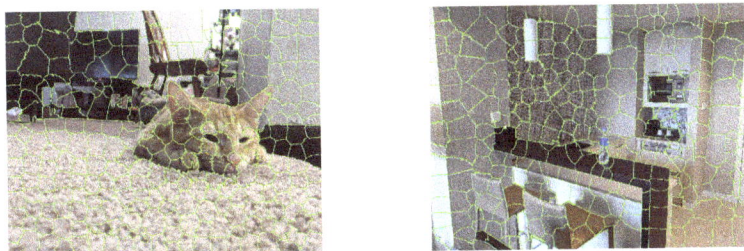

Fig. 8.5. Two examples of superpixel segmentations.

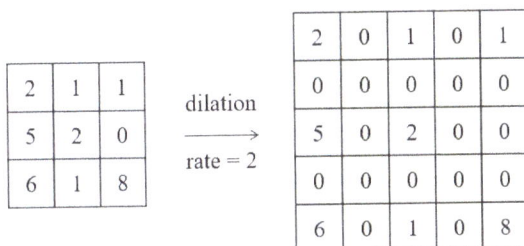

Fig. 8.6. A kernel dilated with a rate $= 2$.

based on superpixels is by Kae *et al.* (2013), who combined conditional random fields (CRFs) and restricted Boltzmann machines (RBMs) to perform image segmentation based on superpixels. We also briefly introduce CRFs later in the book.

We introduce two popular CNN-based segmentation models here, namely DeepLab (Chen *et al.*, 2017a) and U-Net (Ronneberger *et al.*, 2015).

As the word "deep" in its name suggests, DeepLab is a method based on deep neural networks, and mainly CNNs. It seems that "lab" refers to the fact that this model is developed from lab research and is evolving. Indeed, similar to YOLO, DeepLab also has several versions. The very first version of DeepLab introduced atrous convolutions for semantic segmentation. Atrous convolution is also known as dilated convolution. The specific operation is to expand the convolution kernel with holes (zeros) between the entries. For instance, Figure 8.6 shows a dilation with rate 2. When using dilated convolution with a large rate, the convolution captures information at a large scale, which is useful for more global features. With different rates, dilated convolution extract multi-scale contexts, which are important for semantic segmentation. We remark that the dilation operation is a crucial component of wavelets and multi-resolution analysis (Daubechies, 1992).

DeepLabv2 further improved the architecture with the so-called atrous spatial pyramid pooling (ASPP) to extract multi-scale features. DeepLabv3 incorporated an encoder–decoder architecture and ResNets containing atrous convolutions. The details can be found at http://liangchiehchen.com/projects/DeepLab.html.

In both DeepLabv1 and DeepLabv2, CRF is employed to post-process the results of CNN. Nevertheless, CRF is not used in DeepLabv3 since the overall architecture suffices to provide segmentation results of very high quality.

Pretrained versions of DeepLabv3 (Chen *et al.*, 2017b) can be called from PyTorch using the following code. See https://pytorch. org/hub/pytorch_vision_deeplabv3_resnet101/ for an example:

```python
import torch
model = torch.hub.load('pytorch/vision:v0.10.0',
    'deeplabv3_resnet50', pretrained=True)
# or any of these variants
# model = torch.hub.load('pytorch/vision:v0.10.0',
    'deeplabv3_resnet101', pretrained=True)
# model = torch.hub.load('pytorch/vision:v0.10.0',
    'deeplabv3_mobilenet_v3_large', pretrained=True)
model.eval()
```

Fig. 8.7 shows an example of segmentation output from DeepLab.

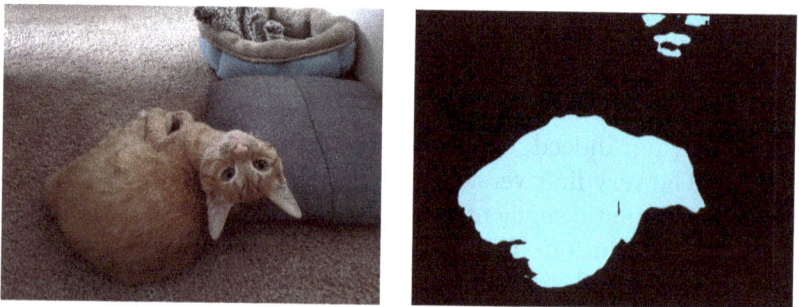

Fig. 8.7. An example of image segmentation using DeepLab.

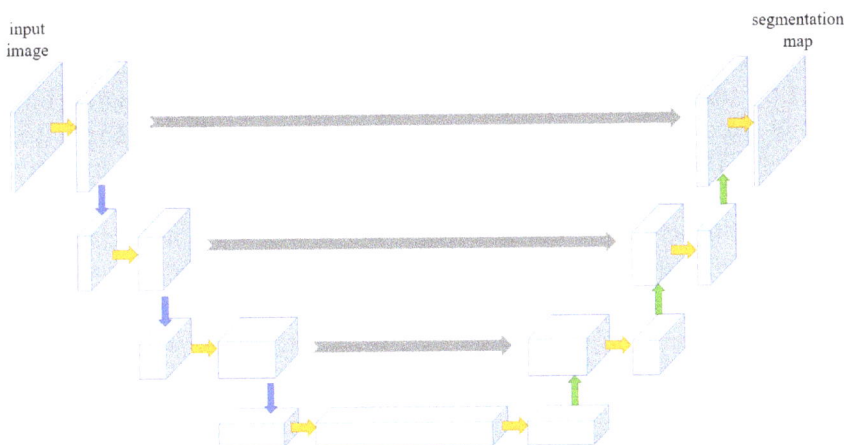

input image

segmentation map

Fig. 8.8. A conceptual illustration of U-Net (simplified). The cuboids stand for the feature tensors; the right arrows in yellow stand for convolutions with activation; the down arrows in blue stand for downsampling operations such as max pooling; the up arrows in green stand for upsampling operations such as deconvolution; the notched arrows in gray stand for copy (and crop to match sizes).

U-Net is another popular CNN model for segmentation tasks. It is especially widely used for medical image analysis such as cell and organ segmentation. It is named according to its U-shape structure. Figure 8.8 serves as a conceptual illustration of U-Net. You may have already noted that, despite the U-shape in the picture, the architecture is very similar to an autoencoder. The main difference is the gray notched arrows, which are skip connections that directly copy features from the encoder. The skip connections preserve details of the images so that they are not lost during the encoding process.

There are many variations of U-Net. Its simple architecture and effectiveness have gained much popularity and it is a baseline in many tasks. It can also be easily adapted with other mechanisms, such as vision transformers (covered in later chapters).

8.3.1 *CNN audio example: U-Net for source separation*

The problem of *source separation* consist of the isolation of individual sounds (or sources) in an audio mixture. CNNs have become a popular approach toward this task, as their convolution and transpose (reverse) convolution operations can be used to generate masks to filter spectral representations of audio into individual sources. The starting point of such process is transforming the audio signal into a time–frequency representation by applying so-called Short Time Fourier Transform (STFT) to the signal. We do not go into the details of this signal processing method, but for our purpose, it is sufficient to say that under broad conditions, such a transformation from time signal to frequency domain is simply a change in representation done by applying a Fourier transform on short segments (windows) in time, sequentially. Each such analysis window changes the representation of the audio signal from a sequence of samples of the audio waveform into a complex vector of frequencies and phases. Repeated application of this process over subsequent segments results in a complex matrix that has the frequency on one dimension and time index corresponding to the segment of the original waveform, as the other. This complex STFT matrix is often represented as amplitude and phase, and commonly the phase part is discarded for the purpose of the source separation analysis, leaving us with a real matrix (practically an image) that shows the distribution of signal energy at different frequencies at different times. This is usually called a magnitude spectrogram but often just spectrogram for short. Other transformations, such as Mel-Frequency analysis or other transformations from the time-domain audio signal to time–frequency matrix are also common. The U-Net analysis of the spectrogram provides the masking of the segments in the time–frequency representation that belong to individual musical instruments or audio sources. Such a *mask* layer is overlaid on the original spectrogram (or mathematically speaking, element-wise multiplied) to allow only the information associated with a particular source to pass through. From this masked spectrogram, the sound of a single source can be approximately reconstructed. To do this, the size of the output of the CNN must match the size of the input (that is, the mask must be the same size as the spectrogram so that it perfectly covers). This naturally requires an encoder–decoder structure to the network so that the learned parameters can be structured in a way that rebuilds the downsampled input to its original size. The U-Net architecture (Ronneberger *et al.*, 2015) is a popular architecture for such problems due to its symmetric U-shape, use of residual "skip connections" to

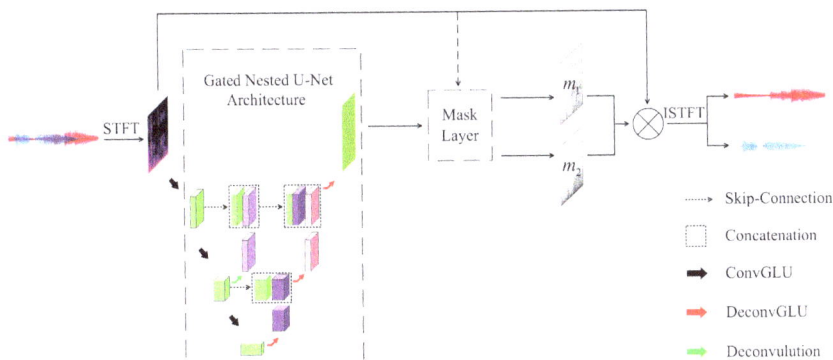

Fig. 8.9. Geng *et al.*'s illustration of the Gated Nested U-Net (Geng *et al.*, 2020); in this use case, the network generates a mask which can be elementwise multiplied with the original audio to separate the sources.

pass information between spatial resolutions, and inclusion of layers of parameters at the decoding side to assist in reconstruction.

An examples of this technique can be found in work by Geng *et al.* (2020) that uses a U-Net variant called "gated nested" U-Net (GNUnet), where there are nested series of layers "filling in" the center of the U. See Figure 8.9 for an illustration of its architecture. The original U "backbone" has gating units applied, a mechanism originally proposed for RNNs which control the information flow throughout the network to allow for modeling more sophisticated interactions. The outputs of GNUNet are used to create a time–frequency spectral mask to generate two masks: one for singing and one for accompaniment. The two masks can be multiplied by the magnitude and phase spectra of the mixture and then transformed back into time-domain signals simultaneously, as opposed to networks which isolate and extract only one source. Also using a U-Net architecture, Kong *et al.* additionally estimated complex ideal ratio masks (that is, decoupling the masks into a mask for magnitude and a mask for phase) to decrease reconstruction error. The result is a system effective at separating vocal tracks, bass, drums, and more.

After application of the mask belonging to the different instruments, the corresponding magnitude and phase are retained and then used to invert the complex spectrogram (STFT) back to time. This recreates the audio signal of the individual sources in the mix, thus accomplishing the sound separation task.

Chapter 9

Recurrent Neural Network

9.1 Sequence Modeling with RNNs

Recurrent Neural Networks (RNNs) is an early neural network application that models time sequences. RNN is learning the probability of a sequence of tokens, $\mathbb{P}(w_1, w_2, \ldots, w_n)$, from a collection of example sequences used in training. To do that, the model keeps a memory for what has happened before each token by maintaining a representation that is coded as an internal state. Overall, RNN can be considered as a function whose output depends on the immediate input as well as its own previous outputs. Unlike traditional feedforward networks, RNNs allow for loops that enable information to persist across time steps. This makes them particularly suited for tasks like language modeling, time-series prediction, and sequential signal analysis.

RNNs are commonly considered as a way to learn a language model, predicting the next token of a sequence given the past, that is, computing $\mathbb{P}(w_n|w_1, w_2, \ldots, w_n)$. This can be considered as a supervised learning problem, in which the sequence $(w_1, w_2, \ldots, w_{n-1})$ acts as an input that is mapped to an output w_n.

Recurrent neural networks can learn long-range influences by encoding the sequential history into a hidden state, which we refer to at time t by $\mathbf{h}_t$. This state $\mathbf{h}_t$ is updated with input $\mathbf{x}_t$ and then used in predicting output $\mathbf{y}_t$. Encasing this history state in a black box, the RNN appears to receive a sequence of vectors $\mathbf{x}_0, \mathbf{x}_1, \ldots, \mathbf{x}_T$ as input and produces a sequence of vectors $\mathbf{y}_0, \mathbf{y}_1, \ldots, \mathbf{y}_T$ as output.

Internally, we start the computation with predicting the next hidden state:

$$\mathbf{h}_t = F(\mathbf{h}_{t-1}, \mathbf{x}_t) \tag{9.1}$$

and then prediction of output using the updated history:

$$\mathbf{y}_t = G(\mathbf{h}_t). \tag{9.2}$$

The relationships between the inputs, multi-dimensional hidden history state, and outputs can be learned with neural network training procedures, as studied in previous chapters. These relationships are represented by functions F and G, which may be composed as

$$F(\mathbf{h}, \mathbf{x}) = A_1(\mathbf{W}_{hh}\mathbf{h} + \mathbf{W}_{xh}\mathbf{x}) \tag{9.3}$$

and

$$G(\mathbf{h}) = A_2(\mathbf{W}_{hy}\mathbf{h}). \tag{9.4}$$

In this notation, A_i stands for an activation function and $\mathbf{W}_{IO}$ stands for a matrix of learnable weights which map from the dimensionality of the matrix's input vector class to the matrix's output vector class. Specifically, $\mathbf{W}_{xh}$ describes the influence of the input on the next state, $\mathbf{W}_{hh}$ describes the influence of the current state on the next state, and $\mathbf{W}_{hy}$ describes the influence of the next state on the next output. An illustration of the relationships between input, hidden state, and output is shown in Figure 9.1.

If we ignore the intermediate outputs $\mathbf{Y}_1, \ldots, \mathbf{Y}_{N-1}$, this is really just a feedforward network, with initial inputs $\mathbf{H}_0$ and $\mathbf{X}_1$, with additional inputs concatenated to the outputs of particular intermediate layers. However, in a traditional feedforward network, the weights between layers are uniquely learned. In the case of an RNN, these weights are *shared*; that is, the same weights will act on the input $\mathbf{X}_t$ regardless of time, and the same weights will act on one state $\mathbf{H}_t$ to create the next (or to create output) regardless of time.

How do we learn in this shared-weight scenario? In our previous backpropagation example, the derivative of loss with respect to a particular weight could be calculated and the weights adjusted

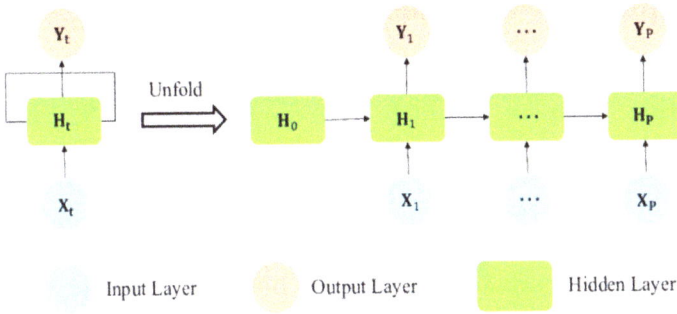

Fig. 9.1. In its compact form (left), RNN consists of a learned hidden state **H**, which is driven by both the current input and past values, and itself drives the output **Y**. To make explicit the development of state **H** over time and its relationship to input and output, we can imagine the RNN in its "unfolded" form (right).

accordingly. Now, for the same weight, we have multiple possible derivatives since there are multiple inputs influenced by that parameter!

Without diving too deep into theory, let's consider an approach by which we take the average of these gradient contributions. That is, for any of the N instances of w_i in the network, which we can call w_{i_n}, we can compute $\frac{\partial L}{\partial w_{i_n}}$. Then, when deciding how to modify shared parameter w_i, we use

$$\frac{\partial L}{\partial w_i} = \frac{1}{N} \sum_{n=1}^{N} \frac{\partial L}{\partial w_{i_n}}. \tag{9.5}$$

In principle, the average should move us closer toward an optimal solution since the largest amount of "mistake" explained by the weight applied at a particular layer will have the largest amount of influence on this averaged gradient component.

There is one critical issue with this approach (and in fact, with any network sufficiently deep), referred to as the *vanishing gradient problem*. This problem is particularly notable for modeling sequential data since it effectively limits how far back in time a model can learn to "reach" to influence its output; while the RNN is capable of representing long sequences of data, it loses the ability to learn at length.

9.2   Input Representation

In order to apply RNN to a real-world data, the choice of input representation is critical. A simple example is a char-RNN, where each character in a text is represented as a one-hot vector. For a vocabulary of size V, each input x_t is a vector of length V where only one element is 1, and the rest are 0s. While effective for small vocabularies, this representation becomes inefficient for larger vocabularies and does not capture any relationship between characters.

To address this, more sophisticated representations like word embeddings are used. Instead of one-hot vectors, each word or character is represented as a dense vector of continuous values, which allows for a more compact and semantically meaningful representation. The embedding vectors can either be pretrained (e.g., using Word2Vec and GloVe) or learned during the RNN training process.

Beyond word embeddings, RNNs are also used with more complex inputs such as the following:

(1) **Text embeddings:** Vectors that represent entire sentences or documents.
(2) **Signal embeddings:** Representing complex signals such as sensor data.
(3) **Time-series embeddings:** Used for capturing temporal patterns in time-series data.
(4) **Audio embeddings:** Used in speech recognition or sound classification tasks.

These embeddings can be learned end-to-end, which allows the model to optimize representations for a specific task, capturing both syntactic and semantic information.

9.2.1   *Learning embeddings with RNNs*

When learning embeddings within an RNN model, the embedding layer maps input tokens (e.g., characters or words) to dense vectors, which are then fed into the RNN. This contrasts with using pretrained embeddings where the vectors are fixed and only the RNN is trained. By learning the embedding, the model can adapt the representation to the task, potentially outperforming fixed embeddings, but it also increases the complexity of training.

Learned vs. Pretrained Embeddings: Learned embeddings are trained alongside the RNN, making them highly task-specific. The embeddings evolve during training to capture patterns relevant to the dataset. However, they require more training data and computational resources to develop a robust representation.

Pretrained embeddings, on the other hand, are generated from large corpora in an unsupervised manner, capturing a wide range of linguistic features. While they can speed up training and provide good general representations, they may not be as effective for domain-specific tasks compared to embeddings learned during training.

Here's a complete example of how to prepare text data for an RNN model in Keras, using a book like Moby Dick as input. The model will train to predict the next character in a sequence, and we will generate text using both the single most probably next prediction using argmax and temperature-based sampling that allows interpreting the prediction vector as probabilities, with the temperature parameter being used in softmax as a way to differently translate the activation functions into probabilities. Temperature of 1 is the standard softmax, while very low temperatures tend to approximate argmax choice, and high temperature to make all choices equal. We see that in the following code:

1. Text Preparation

We start by preparing the text, converting it into sequences of characters that the RNN can learn from:

```python
# Load the text corpus (e.g., Moby Dick)
path_to_file = 'moby_dick.txt'
with open(path_to_file, 'r', encoding='utf-8') as file:
    text = file.read().lower()

# Create a character-level tokenizer
chars = sorted(list(set(text))) # List of unique characters
char_to_index = {char: i for i, char in enumerate(chars)}
index_to_char = {i: char for i, char in enumerate(chars)}
```

```
vocab_size = len(chars)

# Prepare the text sequences for training
sequence_length = 100 # Length of input sequences
sequences = []
next_chars = []

for i in range(0, len(text) - sequence_length):
    sequences.append(text[i: i + sequence_length])
    next_chars.append(text[i + sequence_length])

# Vectorize the sequences and next chars
X = np.zeros((len(sequences), sequence_length), dtype=np.int32)
y = np.zeros((len(sequences), vocab_size), dtype=np.float32)

for i, seq in enumerate(sequences):
    X[i] = [char_to_index[char] for char in seq]
    y[i, char_to_index[next_chars[i]]] = 1 # One-hot encode the next
        char

# Shape of X: (num_sequences, sequence_length)
# Shape of y: (num_sequences, vocab_size)
```

For convenience and reproducibility, here is also a small code for downloading Moby Dick from the web:

```
import requests

# URL to the plain text version of Moby Dick
url = 'https://www.gutenberg.org/files/2701/2701-0.txt'

# Download the text
response = requests.get(url)
text = response.text.lower() # Convert to lowercase for consistency

# Optional: Save the text to a file (for future use)
with open('moby_dick.txt', 'w', encoding='utf-8') as file:
    file.write(text)

# Print the first 1000 characters to verify it was downloaded
    correctly
print(text[:1000])
```

Next comes the Model Definition: We define an RNN model with an embedding layer to learn character embeddings:

```
embedding_dim = 64 # Embedding dimensionality

# Build the model
model = Sequential()
model.add(Embedding(input_dim=vocab_size, output_dim=embedding_dim,
    input_length=sequence_length))
model.add(SimpleRNN(128, activation='tanh'))
model.add(Dense(vocab_size, activation='softmax')) # Output layer
    predicts next character

# Compile the model
model.compile(optimizer='adam', loss='categorical_crossentropy')

# Summary of the model
model.summary()
```

Training the model:

Train the model using the text data, for example, with 20 epochs. Training data is sequences of characters, and the model is trained to predict the next character in the sequence:

```
# Train the model
model.fit(X, y, batch_size=128, epochs=20)
```

Text generation with RNN:

We generate text by feeding an initial seed and predicting subsequent characters. We use both argmax (greedy) sampling and temperature-based sampling:

```
import numpy as np
import random
import sys

# Helper function to sample an index from a probability array
def sample(preds, temperature=1.0):
    preds = np.asarray(preds).astype('float64')
    preds = np.log(preds + 1e-8) / temperature
    exp_preds = np.exp(preds)
```

```
    preds = exp_preds / np.sum(exp_preds)
    probas = np.random.multinomial(1, preds, 1)
    return np.argmax(probas)

# Text generation function
def generate_text(model, seed, num_chars, temperature=1.0):
    generated_text = seed
    for _ in range(num_chars):
        # Vectorize the seed sequence
        input_seq = np.zeros((1, sequence_length))
        for t, char in enumerate(seed):
            input_seq[0, t] = char_to_index[char]

        # Predict the next character probabilities
        preds = model.predict(input_seq, verbose=0)[0]

        # Choose next character based on temperature
        if temperature == 0.0: # Greedy sampling
            next_index = np.argmax(preds)
        else: # Temperature-based sampling
            next_index = sample(preds, temperature)

        next_char = index_to_char[next_index]

        # Add the next character to the generated text
        generated_text += next_char

        # Update seed for the next iteration
        seed = seed[1:] + next_char

    return generated_text
```

Argmax vs. temperature-based sampling:

Argmax sampling always picks the character with the highest probability. This can lead to deterministic and repetitive text generation because it lacks variability.

Temperature-based sampling allows for more diversity by tuning the randomness of predictions. Higher temperatures (>1.0) make the model more "creative" but potentially produce nonsensical text.

Lower temperatures (<1.0) make the model more deterministic, similar to greedy sampling:

```
# Generate text using temperature-based sampling
temperature = 0.8 # Adjust this value to explore different behaviors
generated = generate_text(model, seed_text, num_chars=500,
    temperature=temperature)
print(f"Generated text (Temperature Sampling, temp={temperature}):")
print(generated)
```

Those going carefully over the code must have noted that in the model definition code we manually created a character-level tokenizer with char_to_index and index_to_char dictionaries. Keras has a pre-built Tokenizer module that could be used in the context of text processing and neural networks.

Tokenizer (Keras): The Tokenizer is a Keras utility class typically used to vectorize a text corpus, converting words or characters into sequences of integer indices. It helps convert text into a format that can be used as input to a neural network. Tokenizer can be used for word-level or character-level tokenization. For instance, in word-level tokenization, each unique word is mapped to an integer.

To obtain a one-hot encoding, to_categorical function can be used to convert integer labels into one-hot encoded vectors. When the model is predicting one of several classes (e.g., the next character in a sequence), each class can be represented as a one-hot vector.

Instead of learning the embedding from scratch, you can use pre-trained text embeddings like Word2Vec (Mikolov *et al.*, 2013) or GloVe (Pennington *et al.*, 2014) to initialize the embedding layer in your RNN model. In a good embedding, words which are semantically similar will be closer to each other within the vector space (that is, the distance between their vectors will be minimal). This is motivated by the distributional hypothesis from the field of linguistics: "Words that are used and occur in the same contexts tend to purport similar meanings" (Harris, 1954). Using pretrained embeddings is common in NLP tasks, as pretrained embeddings often capture a rich representation of words based on vast corpora, allowing the model to learn better.

To use a pretrained embedding, you can download pretrained GloVe embeddings (say, the 100-dimensional version trained on 6B tokens). The next step is to prepare the Embedding Matrix for each word in your corpus, retrieve its corresponding GloVe vector, and initialize the embedding layer with these vectors. Finally, use the embedding matrix to initialize the embedding layer and build the rest of the RNN model. If you prefer Word2Vec embeddings, the procedure is similar, where you'd need to load Word2Vec embeddings (which can be obtained using Gensim) instead of GloVe. You can also set the corresponding flags in the call (trainable=True) if you want to fine-tune the embeddings during training, though this increases the risk of overfitting for small datasets.

The word2vec model was applied to music by Herremans and Chuan (2017), showing that the embedded vector space captures tonal relationships from only the context of a musical slice (that is, without any explicit information about the pitch, duration, or intervals contained in the slice).

9.3   GRU and LSTM

In order to mitigate the difficulties of training RNN for long sequences, several models have been developed, including the *Gated Recurrent Unit* (GRU) and the *Long Short-Term Memory* (LSTM) network. We focus on the introduction of the GRU.

9.3.1   *Introduction to GRU*

At a high level, the GRU works in the same fashion as a basic RNN. The input is a sequence of vectors presented one by one. The GRU maintains state, capturing some key aspect of what it has seen so far. This state combined with the next input determines the next state. The advancement comes in the state transition function.

GRU is a simpler architecture than LSTM but still effective in managing long-term dependencies in sequence data. GRUs introduce gates that control the flow of information, deciding how much of the past information to retain and how much of the new input to accept. Specifically, GRUs have two gates:

Update gate: This controls how much of the previous hidden state to carry forward.

Reset gate: This decides how much of the past information to forget. GRU equations.

Let's break down the core operations of the GRU. Assume the input at time step t is x_t and the hidden state is h_t.

Reset gate:

$$r_t = \sigma(W_r \cdot [h_{t-1}, x_t]).$$

The reset gate decides how much of the previous hidden state to forget when generating the new hidden state.

Update gate:

$$z_t = \sigma(W_z \cdot [h_{t-1}, x_t]).$$

The update gate controls how much of the previous hidden state should be carried forward.

Candidate hidden state:

$$\tilde{h}_t = \tanh(W_h \cdot (r_t \odot h_{t-1}, x_t)).$$

This is the new candidate hidden state, which depends on the reset gate and combines the previous hidden state and current input.

Final hidden state:

$$h_t = (1 - z_t) \odot h_{t-1} + z_t \odot \tilde{h}_t.$$

The final hidden state is a blend of the previous hidden state and the new candidate hidden state, controlled by the update gate. In the GRU equations, the small dotted circle symbol $\odot$ represents the element-wise (Hadamard) product. This operation takes two vectors or matrices of the same shape and multiplies their corresponding elements without summation like what is done in a dot product, thus leaving the output in the same dimension as the input vectors.

Key Insights of GRU are as follows: Update Gate (z) behaves somewhat like a memory filter. It decides how much of the past information to retain. If the update gate is close to 1, most of the information from the past is carried forward.

Reset Gate (r): This gate helps reset the memory. When the reset gate is close to 0, the past hidden state is ignored, and the model focuses on the new input.

It should be noted that in a GRU (Gated Recurrent Unit), the reset and update gates perform element-wise operations on the hidden state. This allows different parts of the hidden state to focus on different aspects of the sequence, learning to remember or forget information over different time scales. According to this intuition, each dimension of the hidden state in a GRU can encode different types of information or memory spans because element-wise gates (reset and update gates) operate independently on each dimension of the hidden state. For instance, in a sequence of text or time-series data, one dimension of the hidden state could capture short-term dependencies (e.g., recent words or recent time steps), while another dimension could track long-term dependencies (e.g., over-arching trends or distant events in the past). By controlling each dimension separately, the GRU can selectively forget (if the Reset Gate for that dimension is near 0) or update (if the Reset Gate is near 1) different aspects of the hidden state based on the context at each time step. Relatedly, the role of the Update Gate is favoring new Information. When the Update Gate is near 1 for a particular hidden state dimension, the GRU prioritizes recent information for that dimension. When it's near 0, it retains long-term information. For example, in a conversation, recent words are often more important than words from far back in the sentence. In a time series, recent measurements often have more predictive power than data from many steps back, or vice versa, the past might be a stronger predictor that few recent data samples. The balance between past and present information is what gives GRUs flexibility.

For example, in text generation, when encountering punctuation like a period, the reset gate might learn to forget the previous context, as the sentence has ended and the model should focus on the new sentence. In time-series forecasting, if a sudden shift or event occurs (e.g., a stock market crash), the reset gate may learn to reset memory, as older data might no longer be relevant to the new conditions. This selective forgetting is what enables GRUs to dynamically adjust to sequences with varying amounts of relevant information in the past. Since the gates operate independently on each hidden state dimension, the GRU can track dependencies at multiple time scales simultaneously.

9.3.2 *Introduction to LSTM*

LSTMs are designed to overcome the limitations of RNNs by incorporating additional gates to regulate the flow of information more precisely. Unlike GRUs, which have two gates, LSTMs have three gates:

Forget Gate: This controls which information to forget from the previous state.

Input Gate: This controls what new information to add to the cell state.

Output Gate: This determines the next hidden state based on the cell state.

LSTM equations are given as follows:

Forget gate:

$$f_t = \sigma(W_f \cdot [h_{t-1}, x_t] + b_f).$$

Input gate:

$$i_t = \sigma(W_i \cdot [h_{t-1}, x_t] + b_i).$$

Candidate cell state:

$$\tilde{c}_t = \tanh(W_c \cdot [h_{t-1}, x_t] + b_c).$$

Cell state:

$$c_t = f_t \odot c_{t-1} + i_t \odot \tilde{c}_t.$$

Output gate:

$$o_t = \sigma(W_o \cdot [h_{t-1}, x_t] + b_o).$$

Final hidden state:

$$h_t = o_t \odot \tanh(c_t).$$

In summary, we might make the following comparison between RNN, GRU, and LSTM:

RNN: A basic RNN maintains a simple hidden state that is updated at each time step. However, it struggles with long-term dependencies because the gradients can either vanish or explode, making it difficult to propagate information over many time steps.

GRU: GRU improves upon the RNN by introducing update and reset gates. It uses the update gate to control how much of the hidden state is carried forward, enabling better handling of long-term dependencies.

LSTM: LSTM is more complex and introduces a cell state in addition to the hidden state. Its three gates — forget, input, and output — allow it to finely control what information is retained, added, or passed through, making it highly effective at learning long-term dependencies.

9.3.3   *GRU and the vanishing gradient problem*

One of the main achievemnts of GRU and LSTM models is that they mitigate the *vanishing gradient problem* through their gating mechanisms (reset and update gates). Here, we describe the mathematical principles that explain why GRUs control the gradients more effectively than standard RNNs.

9.4   Gradient Propagation in Recurrent Neural Networks (RNNs)

In a standard RNN, the hidden state at time step t is updated as follows:

$$h_t = \tanh(W_h \cdot h_{t-1} + W_x \cdot x_t + b_h).$$

Here, W_h and W_x are the weight matrices for the hidden state and input, respectively, and b_h is the bias. The function tanh is used as a nonlinear activation function.

When using Backpropagation Through Time (BPTT), we compute the gradient of a loss function L with respect to the parameters of the network. Specifically, we want to compute $\frac{\partial L}{\partial W_h}$, $\frac{\partial L}{\partial W_x}$, and $\frac{\partial L}{\partial b_h}$. To do this, we need to compute the gradient of the loss with respect to the hidden states and propagate these gradients backward through time.

The gradient of the loss L with respect to the hidden state at time t, denoted as $\frac{\partial L}{\partial h_t}$, can be propagated back to earlier time steps using

$$\frac{\partial L}{\partial h_{t-1}} = \frac{\partial L}{\partial h_t} \cdot \frac{\partial h_t}{\partial h_{t-1}}.$$

The derivative of the hidden state h_t with respect to h_{t-1} is given by

$$\frac{\partial h_t}{\partial h_{t-1}} = W_h \cdot \text{diag}(1 - \tanh^2(W_h \cdot h_{t-1} + W_x \cdot x_t + b_h)).$$

When we propagate the gradient back through T time steps, the total gradient of the loss with respect to h_{t-T} is obtained by repeatedly applying the chain rule:

$$\frac{\partial L}{\partial h_{t-T}} = \left(\prod_{k=0}^{T-1} \frac{\partial h_{t-k}}{\partial h_{t-k-1}} \right) \cdot \frac{\partial L}{\partial h_t}.$$

This product involves repeatedly multiplying terms of the form $W_h \cdot \text{diag}(1 - \tanh^2(h_{t-k-1}))$. Over many time steps, this repeated multiplication can lead to either very large or very small values, depending on the properties of the matrix W_h.

Analyzing the term $\frac{\partial h_t}{\partial h_{t-1}}$ we see that it directly reflects how information (and thus gradients) flows through the hidden states. If the gradient with respect to the hidden state vanishes or explodes, this will have a direct impact on the gradient with respect to the weights because $\frac{\partial L}{\partial W_h}$ is influenced by the gradients of the hidden states.

To compute the gradient of the loss with respect to the weight matrix W_h, we use

$$\frac{\partial L}{\partial W_h} = \sum_t \frac{\partial L}{\partial h_t} \cdot \frac{\partial h_t}{\partial W_h}.$$

Here, $\frac{\partial h_t}{\partial W_h}$ involves the hidden state gradients. If the gradients $\frac{\partial L}{\partial h_t}$ become very small or very large (due to vanishing or exploding gradients), then $\frac{\partial L}{\partial W_h}$ will also be correspondingly affected, causing training difficulties.

The repeated multiplication of the weight matrix W_h, along with the nonlinear activation derivative, determines the gradient scaling. If the eigenvalues of W_h are less than 1, the product of many such matrices will tend toward zero, leading to *vanishing gradients*, and if they are greater than 1, the product of many such matrices will grow exponentially, leading to *exploding gradients*.

9.5 Gradient Flow in GRUs

In GRUs, as we discussed above. the hidden state update is controlled by two gates: the *reset gate* r_t and the *update gate* z_t. To derive the BPTT, we rewrite the key GRU equations with separate input and the hidden state weight matrices:

$$z_t = \sigma(W_z \cdot x_t + U_z \cdot h_{t-1} + b_z) \quad \text{(Update gate)},$$

$$r_t = \sigma(W_r \cdot x_t + U_r \cdot h_{t-1} + b_r) \quad \text{(Reset gate)},$$

$$\tilde{h}_t = \tanh(W_h \cdot x_t + U_h \cdot (r_t \odot h_{t-1}) + b_h) \quad \text{(Candidate hidden state)},$$

$$h_t = (1 - z_t) \odot \tilde{h}_t + z_t \odot h_{t-1} \quad \text{(Hidden state update)}.$$

Here, W_z, W_r, W_h are the input weight matrices, U_z, U_r, U_h are the recurrent weight matrices, and b_z, b_r, b_h are the biases. The σ and tanh represent the sigmoid and hyperbolic tangent activation functions, respectively, and $\odot$ denotes element-wise multiplication.

The gradient of the loss L with respect to the model parameters becomes

$$\frac{\partial h_t}{\partial h_{t-1}} = \text{diag}(z_t) + \text{diag}(1 - z_t) \cdot \frac{\partial \tilde{h}_t}{\partial h_{t-1}}$$

where we expanded the derivative by computing the total derivative of the loss with respect to h_t at time t which is given by

$$\frac{\partial L}{\partial h_t} = \frac{\partial L}{\partial h_T} \cdot \prod_{k=t}^{T-1} \frac{\partial h_{k+1}}{\partial h_k}$$

and used the update equation for h_t, the derivative with respect to the previous hidden state h_{t-1} as

$$\frac{\partial h_t}{\partial h_{t-1}} = \frac{\partial}{\partial h_{t-1}}((1 - z_t) \odot \tilde{h}_t + z_t \odot h_{t-1}).$$

The term $\text{diag}(z_t)$ indicates how much of the previous hidden state is directly carried forward, while $\text{diag}(1 - z_t)$ scales the contribution of the new candidate hidden state $\tilde{h}_t$.

To compute $\frac{\partial \tilde{h}_t}{\partial h_{t-1}}$, we use the expression for $\tilde{h}_t$:

$$\frac{\partial \tilde{h}_t}{\partial h_{t-1}} = (\text{diag}(1 - \tanh^2(\tilde{h}_t))) \cdot U_h \cdot \text{diag}(r_t).$$

Here, $\text{diag}(1 - \tanh^2(\tilde{h}_t))$ is the derivative of the tanh function, U_h is the weight matrix, and $\text{diag}(r_t)$ is the reset gate that modulates how much of the previous hidden state influences the candidate hidden state.

Combining these results, we get

$$\frac{\partial h_t}{\partial h_{t-1}} = \text{diag}(z_t) + \text{diag}(1 - z_t) \cdot (\text{diag}(1 - \tanh^2(\tilde{h}_t)) \cdot U_h \cdot \text{diag}(r_t)).$$

In the expression for $\frac{\partial h_t}{\partial h_{t-1}}$, the update gate z_t controls the extent to which the gradient flows back unchanged. If $z_t \approx 1$, the term $\text{diag}(z_t) \approx I$, and the gradient will pass through without much attenuation. The factor $1 - z_t$ ensures that, when z_t is small, the model is more likely to allow new information from the current time step to influence the hidden state. This mechanism helps prevent the gradient from vanishing too quickly.

Finally, the gradient of the loss with respect to the parameters (like U_h) involves summing over all time steps:

$$\frac{\partial L}{\partial U_h} = \sum_t \frac{\partial L}{\partial h_t} \cdot \frac{\partial h_t}{\partial U_h}.$$

Since $\frac{\partial h_t}{\partial U_h}$ depends on the gradients of the hidden states, the gating mechanisms (update and reset gates) influence the stability of these gradients. By modulating how much information is passed through, the gates effectively regulate the contribution to the parameter gradients, helping control both vanishing and exploding gradients during training.

It should be noted that while the GRU design can mitigate vanishing gradients through the gating mechanism, the problem of exploding gradients still arises when the weight matrices U_h have large eigenvalues. The term $U_h \cdot \text{diag}(r_t)$ can lead to large gradients if the values in U_h are large or if the reset gate r_t is mostly active (close to 1).

To control exploding gradients, techniques such as

- **gradient clipping** (limiting the norm of the gradients),
- **regularization** (like weight decay), or
- **orthogonal initialization** of weight matrices

are commonly used in practice.

Finally, for the sake of completeness, we will summarize the BPTT effect on vanishing gradients for LSTM as well.

9.6   Vanishing Gradients in LSTM and BPTT Solution

The main LSTM equations are rewritten with separate input and recurrent weight matrices as follows:

$$f_t = \sigma(W_f \cdot x_t + U_f \cdot h_{t-1} + b_f) \quad \text{(Forget gate)},$$

$$i_t = \sigma(W_i \cdot x_t + U_i \cdot h_{t-1} + b_i) \quad \text{(Input gate)},$$

$$o_t = \sigma(W_o \cdot x_t + U_o \cdot h_{t-1} + b_o) \quad \text{(Output gate)},$$

$$\tilde{c}_t = \tanh(W_c \cdot x_t + U_c \cdot h_{t-1} + b_c) \quad \text{(Candidate cell state)},$$

$$c_t = f_t \odot c_{t-1} + i_t \odot \tilde{c}_t \quad \text{(Cell state update)},$$

$$h_t = o_t \odot \tanh(c_t) \quad \text{(Hidden state update)}.$$

Here, W_f, W_i, W_o, W_c are the input weight matrices, U_f, U_i, U_o, U_c are the recurrent weight matrices and b_f, b_i, b_o, b_c are biases. The forget gate f_t, input gate i_t, and output gate o_t control how information flows through the memory cells.

The derivative of the cell state with respect to its previous value is given by

$$\frac{\partial c_t}{\partial c_{t-1}} = f_t.$$

Since f_t is controlled by the forget gate, it can adjust how much of the previous cell state is carried forward. When $f_t \approx 1$, the gradient will pass through without significant attenuation, preventing vanishing gradients. This design helps maintain gradients over long sequences through the following mechanisms:

- **Memory cell c_t:** The cell state updates are additive, which prevents the gradients from diminishing too quickly since the forget gate f_t allows controlling the scale of gradient flow.
- **Forget gate f_t:** The forget gate modulates how much past information is retained, directly influencing gradient propagation.
- **Output gate o_t:** By scaling the output, the LSTM controls when the hidden state should be influenced by the memory cell.

While LSTM can manage vanishing gradients through the gating mechanisms, exploding gradients may still occur if the forget gate f_t remains consistently large. To address this, techniques like gradient clipping are often employed to keep gradients within a manageable range.

9.7 Stacked RNNs

A stacked RNN simply adds one or more hidden layers to this computation. That is, the state update function becomes "deep". This makes the RNN architecture more versatile and richer, considering temporal dynamics of the hidden state as well as the dynamics of the data as well, possibly repeated over several layers. Consider the state update function of an RNN:

$$\mathbf{h}(t) = f(\mathbf{h}(t-1), \mathbf{x}(t)). \tag{9.6}$$

In the ones we have seen so far (the basic RNN and the GRU),

$$f(\mathbf{h}, \mathbf{x}) = S_1(\mathbf{W}_{hh}\mathbf{h} + \mathbf{W}_{xh}\mathbf{x}). \tag{9.7}$$

We may read this as "$\mathbf{h}(t)$ is obtained from $\mathbf{h}(t-1)$ plus $\mathbf{x}(t)$ in a feedforward architecture with zero hidden layers".

This gives L functions $f_1, \ldots, f_L$ with their own learnable parameters. We can write this as

$$f(\mathbf{h}(l), \mathbf{h}(l-1), l) = S_1(\mathbf{W}_{hh}(l)\mathbf{h}(l) + \mathbf{W}_{xh}(l)\mathbf{h}(l-1)),$$

$$l = 1, \ldots, L, \tag{9.8}$$

where $\mathbf{h}(0)$ denotes $\mathbf{x}$.

The state representation at time t itself is richer. $\mathbf{h}(t) = (\mathbf{h}(0,t), \mathbf{h}(1,t), \ldots, \mathbf{h}(L,t))$. This may be viewed as a hierarchical representation of state in order of increasing coarseness. We see in the following some other hierarchical representations, that employ also different time resolutions in a conditional manner.

9.8 Inference

We already saw an example use of a trained RNN to generate text that resembles the statistics of the training group. In the most broad sense, we can run a sequence of words $w_1, w_2, \ldots, w_{n-1}$ through a trained RNN and produce a probability distribution over the likely next word. Typically, we want more than this. We want the most likely extension of a certain length k of the given sequence. That is, we want

$$w_n, w_{n+1}, \ldots, w_{n+k-1}$$

$$= \underset{w_n, w_{n+1}, \ldots, w_{n+k-1}}{\arg\max} \mathbb{P}(w_n, w_{n+1}, \ldots, w_{n+k-1} | w_1, w_2, \ldots, w_{n-1}).$$

$$(9.9)$$

This inference problem is called the decoding problem. When k is 1, the solution is just the word with the highest probability in the RNN"s output vector after the RNN has processed the words $w_1, w_2, \ldots, w_{n-1}$ sequentially as input. For the general case of k, the decoding problem is intractable. That is, we cannot hope to find the most likely extension efficiently. So, we resort to heuristic methods.

Greedy decoding: One simple approach to finding a good extension of length k is to start by finding the highest-probability word in the output vector, inputting this word to the RNN (so that it outputs the probability distribution over the next word), finding the highest-probability word again, and repeating the process. This approach is called greedy because it makes the best local decision in every iteration.

This approach can work well when, in every greedy step, adding the highest-probability word in the output vector moves us toward a globally optimal solution. As in the following example.

Example: Consider a language model trained on national, state, and county park names. Consider $x^{(1)} = $ yellowstone. The greedy extension will be $x^{(2)} = $ national.

Not surprisingly, such a happy circumstance is not always the case. As in the following example:

Example: Consider a language model trained from a multi-set of organization names. Examples of organization names are Bank of America, Google, and National Institutes of Health. ... The term multi-set means that organization names may be repeated in this list. Now consider the problem of finding the most popular organization name in this list. The first step of the greedy method would pick bank or the or whichever word occurs most frequently in this list. The most popular organization name (say Google) may not start with this first word.

Beam search in RNN: As mentioned earlier, the optimal decoding for infinite or very long sequences into the future is in general intractable (NP-hard). There are many heuristic approaches that try to do better than greedy decoding, at varying costs of running time and memory requirements. Beam search has turned out to be an especially attractive one in sequence-to-sequence (seq2seq) labeling use cases (especially language translation). So, that is what we describe here.

Beam search keeps the k best prefix sequences. In every iteration, it considers all possible 1-step extensions of these k prefixes, computes their probabilities, and forms a new list of the k best prefix sequences thus far. It keeps going until the goal is reached. In our scenario, this would be when the extended sequence lengths are T. It then pulls out the highest-scoring sequence in its list of k candidates as the final answer.

9.9   Sequence-to-Sequence Modeling with RNNs

Sometimes we seek not to learn only an estimated distribution for a sequential pattern but a mapping from one sequence to another. The most straightforward example is text translation. We let RNN see two sequences: an input in source language, say English, and the output in the target language, say French. In order to perform the translation, the problem is split into modeling of the English sentence, summarizing it into the last state, often called the "thought vector", and then starting a second RNN, or decoding, to produce the French translation. The sequence to sequence problem can also appear as a continuation within the same language or data source. One could think of input as a prompt, in which case the target seqeunce can be a longer answer to the prompt, like what is done in large language models (LLMs). Here we limit the problem to much shorter cases and other use cases, such as style transfer. For example, pretend you are a jazz saxophonist, and you attend a friend's classical piano recital. A striking melody from one of their pieces catches your attention, and you'd like to improvise on it when you practice later that day. In this case, you are taking an input sequence of the melody in a classical context, and translating it to a jazz melody (which may have different styles of ornamentation). Such a problem is solved using sequence-to-sequence (often abbreviated seq2seq) modeling.

Formally, we can describe such problems this way: Given vocabularies V_z and V_x and training data pairs (z_n, x_n) independently and identically distributed from some distribution p, can we estimate the distribution $p(x|z)$?

Framing the above problem in this notation, V_z are the possible notes from the piano, and V_x are the possible notes from the saxophone. We can collect samples of the same melody played once by the classical pianist (z_n) and again by the jazz saxophonist (x_n), and use these to learn a distribution of what we expect the jazz saxophone to sound like (x) for a given piano melody (z). There are numerous other problems we can frame as seq2seq, including the problem of transcription (given an audio sequence, returning the same represented music in symbolic notation).

9.10   RNN Applications

RNNs have been used in many applications for different data types and tasks. RNNs, especially those using LSTM, have been widely used for text generation, such as generating coherent and contextually relevant text in various languages. Character-level text generation for tasks like code, literature, and poetry creation. Enhancing creativity and diversity in generated text uses variational autoencoders (VAEs) in conjunction with RNNs. In text analysis, RNNs excel at capturing context and sentiment in text, making them effective for analyzing customer reviews and social media posts for sentiment classification. Performing document-level sentiment classification by hierarchically analyzing sentiment at sentence level and aggregating for overall sentiment. Enhancing sentiment analysis by incorporating attention mechanisms, enabling models to focus on sentiment-bearing words or phrases.

Machine Translation is maybe the most celebrated use of RNNs where neural networks have shown to significantly improve machine translation quality compared to classical NLP methods. Deep RNNs, like Google Neural Machine Translation (GNMT), enhance translation accuracy and fluency by capturing complex patterns and long-range dependencies. Incorporating subword units, such as Byte-Pair Encoding (BPE), allows RNN-based translation models to handle rare and out-of-vocabulary words effectively. Hybrid models combining RNNs with attention mechanisms, and even incorporating transformer architectures, have achieved state-of-the-art performance in translation tasks.

For Audio Data, RNNs have become central to modern speech recognition systems. Early work explored deep neural networks, including RNNs, for speech-to-text systems, showing their effectiveness in capturing temporal dependencies in audio signals. DeepSpeech, a prominent speech recognition system, utilizes LSTM networks and extensive training data for accurate transcription, even in noisy environments. DeepSpeech2 further improved accuracy by incorporating bidirectional RNNs. RNN-transducer (RNN-T) models offer efficient end-to-end speech recognition by integrating acoustic and language models within a single framework.

Time series data: RNNs' strength in modeling temporal patterns makes them well-suited for various time series forecasting applications. Such applications include Finance, Weather forecasting, consumer demand patterns, and more. In Finance Deep RNNs have demonstrated success in predicting stock returns, often outperforming traditional ML models. By combining RNNs with techniques like CNNs, attention mechanisms, transfer learning, and reinforcement learning further enhancement in forecasting performance can be achieved. RNN can be used for capturing meteorological patterns both for short-term and long-term phenomena, including extreme weather events, forecast energy generation from sources like wind and solar power and optimize energy demand forecasting in smart grids. In commerce and manufacturing, RNNs are used to predict consumer demand patterns, optimization of of supply chains and managing inventory.

Other Modalities and Applications for RNN include analyzing biological sequences like DNA, RNA, and proteins. Gene prediction and protein structure prediction, leveraging RNNs' ability to capture dependencies within sequences. Predicting DNA-binding protein sequences with high accuracy using bidirectional LSTM, aiding in understanding protein–DNA interactions. Protein structure prediction and function annotation, capturing sequential dependencies in amino acid sequences for drug discovery and disease research.

RNNs have been used in Autonomous Driving to process sequential data from sensors for path planning, object detection, and trajectory prediction. RNNs are applied across various fields for anomaly detection. For example, in Cybersecurity, RNNs are used for detecting unexpected bursts in network traffic to identify malicious activities. In Industrial Monitoring, RNNs are used to identify anomalies in multivariate time series data collected from machinery for predictive maintenance. In healthcare, RNN are used for detecting abnormal patterns in physiological signals, such as ECG, for early diagnosis and monitoring of conditions such as cardiac arrhythmias.

For a recent survey of these application, one is referred to the following: Mienye *et al.* (2024).

In creative applications, such as music, RNN has been used to generate both symbolic and audio signals. Performance RNN (Simon and Oore, 2017) is a network which generates performance-quality MIDI piano output. You can learn more about Performance RNN on the Google Magenta's blog (https://magenta.tensorflow.org/perfor mance-rnn). We will describe it in more detail next.

9.10.1 *Performance RNN for music generation*

Performance RNN is a machine learning model primarily used to generate expressive piano performances using a symbolic representation rather than audio. The model is a long short-term memory (LSTM)-based recurrent neural network that models polyphonic music, which means it can handle multiple notes played simultaneously. It generates performances with expressive timing and dynamics. You can run the code yourself with instructions available in their GitHub

Repository (https://github.com/magenta/magenta/tree/main/mag enta/models/performance_rnn). What is special about this application, is the complex representation of music data, including multiple co-occurring notes of different durations, an including performance aspects of dynamics, pressing of the piano pedal, and additional conditioning information that can be derived from the data headers or some music analysis. The sequential music data comprising of the following elements is derived from Musical Instruments Digital Interface (MIDI) standard:

- **Note-on events:** Indicating when a note starts.
- **Note-off events:** Indicating when a note ends.
- **Velocity changes:** To capture the dynamics (loudness) of notes.
- **Time-shift events:** Representing the amount of time elapsed between events.
- **Control changes:** Such as pedal controls.

For example, if there are 128 possible note velocities, 88 piano keys (note-on/note-off for each key), and several time-shift increments, the total vocabulary size for the model would be the sum of these categories. RNN uses a special MIDI-Like tokenization scheme with several particular components:

- **Data:** Performance RNN is trained on the Yamaha e-Piano Competition dataset (introduced in Appendix D). This is a collection of approximately 1,400 MIDI files, generated by recording the key-presses of highly skilled classical pianists. Accordingly, the music is expressive in both timing (rubato) and velocity (dynamics).
- **Tokenization:** MIDI-like tokenization is employed. At each step, the input to the RNN is a single one-hot 413-dimensional vector representing the 413 possible note-on, note-off, velocity, and time-shift events.
- **Embedding:** Performances are encoded as sequences of these events and fed into the network using a one-hot vector encoding at each step.
- **Teacher forcing:** The model is trained on 30-second segments from MIDI performances using teacher forcing. This technique involves feeding the actual, correct output from the training data back into the model as input during training, helping it learn the desired sequences more effectively.

- **Model architecture:** Three hidden layers of LSTMs with 512 cells.
- **Training and loss:** The model is trained using the RNN technique of *teacher forcing*. *Teacher forcing* helps models reach stable convergence during training by providing the correct (rather than predicted) output as input to each sequential step. While this is beneficial to efficiently learn the correct patterns, it reduces the model's ability to learn from variations which may not be part of the training data (but may exist otherwise). For generative models which seek to perform and innovate (rather than replicate), this might limit the creativity, while retaining more regularity specific to the musical style or music rules. Log loss (i.e., categorical cross-entropy) is the loss function used to drive training backpropagation.
- **Output generation:** Beam Search is an improvement on greedy search, where the model considers multiple possible output sequences (beams) and chooses the sequence with the highest joint log likelihood. The sources utilize a beam search with a branch factor of 2 (fbeam = 2), meaning each beam generates 2 copies at each step. The number of steps for which outputs are generated is 240 (nsteps = 240), which is approximately 6 s of performance. The number of beams kept in memory is 8 (nbeam = 8).
- **Stochastic beam search:** To avoid low-entropy outputs and repetitions, the sources employ stochastic beam search with a temperature of 1. This method introduces randomness by selecting beams to retain based on probabilities derived from their log-likelihoods.
- **Conditioning:** One limitation of the standard Performance RNN is its lack of control over what it generates. Conditioning provides the user with external controls to influence the music generation process. Conditioning involves feeding additional feature vectors (control signals) to the model alongside the musical events. These control signals can represent various musical aspects like composer, tempo, or velocity. The article explored several types of conditioning, such as Individual Composers or groups of similar composers, Time Period, and Geographical Conditioning that takes into account the composer's location and birth year to reflect regional influences on music. Other conditioning includes information like key, tempo, and musical form the was derived from performance titles in order to guide generation, major and minor musical

scales to control the tonality of the output, tempo keywords from titles into groups, enabling control over the speed of the generated music, and velocity conditioning to influence the perceived loudness and dynamic variations in the output. One particular conditioning related to musical form is relative position that provides information about the position of a 30-second excerpt within the original piece, aiming to capture characteristics of beginnings, endings, or climaxes.

One of the challenges in this work is the sparsity of data of control parameters available for training, which emphasizes the need to find a balance between overfitting and artistic expression. The authors argue that some degree of overfitting might be acceptable in a creative context, as it could lead to "quoting" or stylistic borrowing, similar to human musical practices.

Chapter 10

Attention and Transformers

10.1 Transformers

Transformers have become a popular tool to use in machine learning problems involving text and language for their strengths in sequence modeling and sequence-to-sequence prediction. In this chapter, we see that transformers are similarly well suited to these problems in music and audio.

An important preliminary to understanding transformers is the idea of *embedding*. We already encountered embedding in the RNN chapter. As a reminder, embedding is a representation of some word in the possible vocabulary by creating a vector with weights that, when considering all possible words, represent some information about the contextual properties most regularly found around the word. This context is not encoded in former approaches such as bag-of-words, in which the grammar and usage around the words is lost. This is analogous to viewing an image as a histogram of its pixel intensities, while it still contains some useful information about the contents of the image, most of its useful structure and relationships between pixels are lost.

In the RNN chapters, we described a way to embed information that captured the "meaning" of a token through its relationship to other tokens. However, there is another kind of embedding critical to the transformer: *positional* embedding. In positional embedding, we represent not the meaning of a token but rather the token's position in a sequence. These can either be learned embeddings, or deterministically defined. In the transformer, these two embeddings (meaning and position) are combined to make a complete token embedding. Recall one of the challenges associated with RNNs: Information from early in the sequence can become lost as the sequence progresses, but this information is often still relevant and important downstream. We discussed some architectural changes to RNN that better preserve this information, and this chapter expands on an approach which has exploded in recent years, namely *attention*.

While the RNN variants focused on creating better hidden states which maintain long history, the attention mechanism learns to attend to previous hidden states when relevant. Early attention models appeared in RNN sequence to sequence models, mostly to deal with permutations due to different ordering of words in similar expressions between languages in translation tasks. Longer term attention is required to deal with structural characteristics of text, such as continuations or response to prompts in language models. Another examples is importance of attention for composing music. Musical form takes shape often at both a macro and micro level. A piece might change tonalities in difference sections, also called harmonic modulation (for example, a piece might begin in C major, then modulate to A minor, which is the relative minor, and perhaps to G major or key of the dominant, then back home to C major for conclusion). Within each of these sections, there is a wealth of melodic, harmonic, and rhythmic figures and repetitions, and as a listener we maintain a sense of the long-term structure which gives us that sense of tensions and resolution, often leading to conclusion and closure when we return to the original key. As such, it is important for a neural model to maintain representation of these components simultaneously so that it can learn to create similar complete structure.

Essentially, using attention the model creates a contextual vector composed of previous hidden states, weighed according to their learned relevance to the current task. We can imagine attention to be the mechanism that guides the model to its most relevant memories at a given time. We skip the RNN attention and discuss directly the attention in Transformers. There are several excellent webapps that provide visualization for transformers. Transformer Explainer (Cho *et al.*, 2024) is an interactive visualization tool that can be accessed at https://poloclub.github.io/transformer-explainer/. It runs a live GPT-2 instance in the user's browser, allowing experimenting with text input continuation and displaying its internal components and parameters. Music transformer visualization (Huang *et al.*, 2018b), with comparison and contrast between regular attention and relative attention (Shaw *et al.*, 2018), is available at https://storage.google apis.com/nips-workshop-visualization/index.html.

Self-attention: The self-attention mechanism is a key defining characteristic of transformer models which allows the network to make use of preceding or surrounding tokens to model the sequence distribution. There are three keywords that we use to describe the attention mechanism:

(1) **Query:** The current token (that is, the one being predicted) is used to generate a query.
(2) **Key:** All tokens in the contextual scope map to a key. The "query" from above is "asked" over all of these "keys".
(3) **Value:** Each token in the contextual scope has some associated value vector. The dot product of the query and the key then tell us how much of this value to pass forward in the network.

Taken together, we can consider the inner product of a query with a key to be the severity by which the contextual token deserves attention in making the prediction associated with the current key. The value is then the information associated with that context, which is propagated forward only to the extent of the "attention" granted to it by the query-key pair.

We will adopt the notation of Tay *et al.* (2022) to formally describe in a series of equations the Transformer architecture as explained above, with particular emphasis on multi-head attention. First, we generate a query as

$$q = W_q e + b_q, \tag{10.1}$$

where e is the vectorized version of the current input token, W_q and b_q (and subsequent W and b) are learned weights and biases, and q is the query generated from the token e.

After generating a query, we must generate a set of keys to each meld with the query. A key k_t is generated for each contextual vector e_t:

$$k_t = W_k e_t + b_k. \tag{10.2}$$

Each combined query + key interaction tells the transformer model how much of the *value* associated with the key's token to extract. These values v_t are defined by

$$v_t = W_v e_t + b_v, \tag{10.3}$$

and taken together, the partial value α extracted from each value v_t is determined by the keys as

$$\alpha_t = \frac{e^{q^\top k_t}}{\sum_{u=0}^{T} e^{q^\top k_u}}. \tag{10.4}$$

The dot product between query q and key k is considered an attention matrix, which is responsible for learning alignment scores between tokens within the sequence — that is, it informs the model how a token should gather information from other tokens.[1]

The combined value is then

$$v = \sum_{k=0}^{T} \alpha_t v_t. \tag{10.5}$$

The attention mechanism can be imagined as a learned bias used to pool all tokens based on their relevance. However, this learned model comes at a cost: Self-attention faces scaling issues due to its quadratic time and memory complexity.

The outputs v of each head can be passed into a dense layer to compute output as a combination of values of attention. To summarize this process, for any reference token, different attention heads (since we are using *multi-head attention*) learn to gather information from surrounding tokens in different ways. This information can then be combined in a dense layer, leveraging multiple learned patterns of attention for the task at hand.

With the attention module complete, standard neural network layers can now be employed, such as feedforward layers and activations. For example, Vaswani *et al.* used two feedforward layers with ReLU activations in their sequence-to-sequence application.

---

[1]Note that there is typically a scaling term in the numerator and denominator exponents based on the Transformer dimensionality, which we have omitted for illustrative purposes.

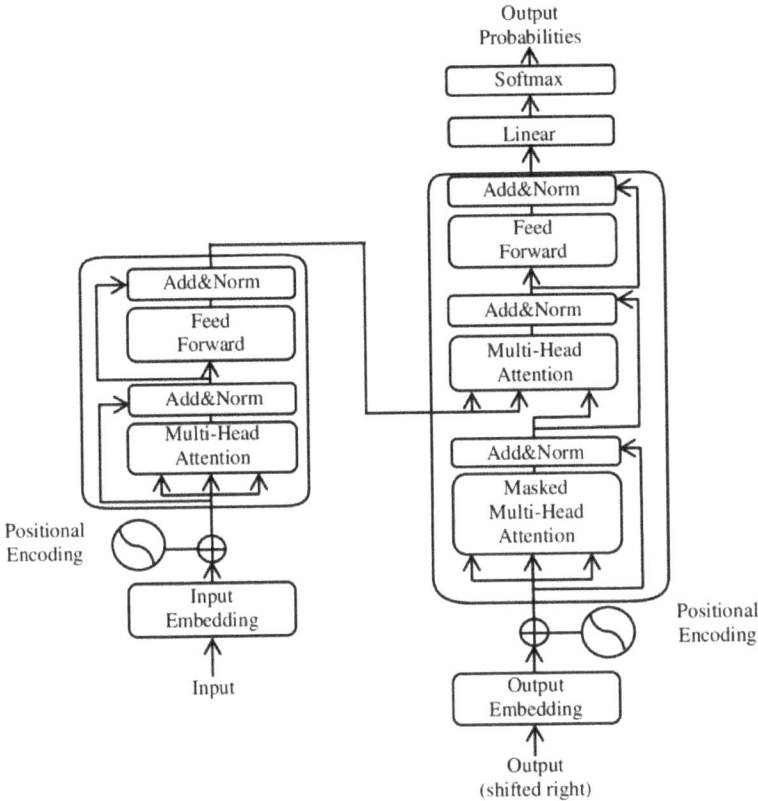

Fig. 10.1. The left segment of this figure shows the attention mechanism described above, and its use within the original transformer architecture proposed by Vaswani *et al.* in 2017. We note that the inputs and outputs of each multi-headed self-attention module are connected *residually*. *Residual connections* describe neural network layers which combine information from earlier layers directly with the output of layers deeper within the network.

10.1.1 *Transformer encoders and decoders*

We previously introduced encoders and decoders as a means of reducing the representation of input data to (and reconstructing data from) a latent, lower-dimensional encoding. Transformers can be thought of in a similar fashion; we are trying to model some distribution $p(y|x)$, and we make an assumption that there exists some parameterization θ which through a learning or optimization process can be used to bring out features in x that the model should attend to, forming a latent space $z = f_\theta(x)$. So, we learn encoding $p(y|z)$. In the case of sequential data, the encoder maps a sequence of observations $(x_1, x_2, \ldots, x_n)$ to a sequence of latent representations $(z_1, z_2, \ldots, z_n)$, which are passed to the decoder that generates an output sequence $(y_1, y_2, \ldots, y_m)$. One should note that the length of the output sequence can be different from the length of the input sequence.

In the original formulation of the transformer (Vaswani *et al.*, 2017), shown in Figure 10.1, was used in two fashions: as a way for the decoder to attend to the input, similar to attention in RNN, and as an intra-attention, or self-attention, by letting the input sequence "listen to itself", thus capturing the dependencies across all positions of the input sequences simultaneous rather then sequentially. Compared to RNN, where the number of operations grows with the length of the input data and is impossible to parallelize, in the Transformer the number of operations is constant, albeit at the cost of reduced resolution due to averaging by the attention-weighted positions. This effect was counteracted by using Multi-Head Attention, basically replicating the self-attention with multiple learned weights, which amounts to extracting multiple features in parallel.

In RNN with attention as well as in Transformer, the mapping between input an output is done by an inter- or cross-attention mechanism as depicted in Figure 10.2. The cross-attention mechanism in Transformer architecture mixes the features from x and y by performing a dot-product between their embedding sequences after they were mapped to the same dimension. The output length of the decoder is determined by the target sequence y as it plays a role of a query with input x used to produces key and value. Technically, in cross-attention, the queries are generated from one embedding and keys and values are generated from another embedding.

One more crucial element to our sequence-to-sequence Transformer is positional embedding, introduced earlier in this chapter:

```
class PositionalEmbedding(layers.Layer):
    def __init__(self, sequence_length, vocab_size, embed_dim,
                 **kwargs):
        super().__init__(**kwargs)
        self.token_embeddings = layers.Embedding(
            input_dim=vocab_size, output_dim=embed_dim
        )
```

```
self.position_embeddings = layers.Embedding(
    input_dim=sequence_length, output_dim=embed_dim
)
self.sequence_length = sequence_length
self.vocab_size = vocab_size
self.embed_dim = embed_dim

def call(self, inputs):
    length = tf.shape(inputs)[-1]
    positions = tf.range(start=0, limit=length, delta=1)
    embedded_tokens = self.token_embeddings(inputs)
    embedded_positions = self.position_embeddings(positions)
    return embedded_tokens + embedded_positions

def compute_mask(self, inputs, mask=None):
    return tf.math.not_equal(inputs, 0)
```

This is used to make the model aware of the element order in the sequence. It works by defining a range from 0 to the length of

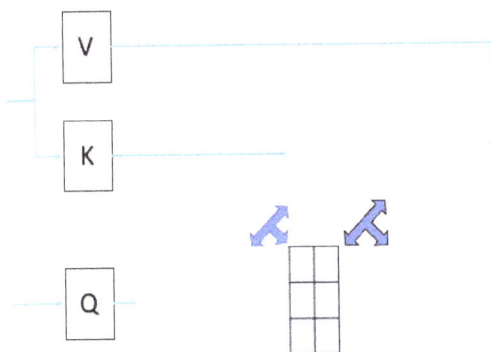

Fig. 10.2. Illustration of the cross-attention mechanism. Some input sequence (shown in gray) is transformed by matrix W_V and W_K. These transformations do not need to preserve the original shape of the input. In this example, W_V must be a 5×2 dimension matrix, acting on the (transposed) 2×3 input sequence, giving (transposed) 5×2 value embeddings. Similarly, the key transformation W_K must be 4×3, giving 4×2 output. A query transformation W_Q is applied to a separate input sequence (shown in blue), in this example, opting to preserve shape. From the dot product of the query and key, we arrive at a matrix of attention scores (3×2). When multiplied by a (2×5) value matrix, we reach our Transformer output, a (3×5) matrix of weighted values from one of the input sequences, with weights determined by both that input and the additional query input, hence attention that "cross"es both inputs.

the input sequence(s), then creating an embedding of the position's index as a vector of dimensionality embed_dim. This can be simply added to the sequence element's token embedding to create a complex representation of both the element's content and position.

We now have everything we need to assemble the model: First, we gather the vectorized sequence elements and then add to these their positional embeddings. We apply the Transformer Encoder to this complex input. Likewise, the decoder receives as inputs the "preceding output" (aka decoder_inputs) and the encoded representation of the sequence inputs. The decoder applies positional embedding and attention to the "preceding output" sequence and then combines this densely with the encoded representation of the sequence inputs, outputting at last an element with the size of the desired vocabulary.

```
embed_dim = em  # These values should be
latent_dim = la #  constants, tailored to
num_heads = nu  #  your problem and data.

encoder_inputs = keras.Input(shape=(None,),
                dtype="int64", name="encoder_inputs")
x = PositionalEmbedding(sequence_length, vocab_size, embed_dim)
                (encoder_inputs)
encoder_outputs = TransformerEncoder(embed_dim, latent_dim,
                                num_heads)(x)
encoder = keras.Model(encoder_inputs, encoder_outputs)

decoder_inputs = keras.Input(shape=(None,), dtype="int64",
                        name="decoder_inputs")
encoded_seq_inputs = keras.Input(shape=(None, embed_dim),
                        name="decoder_state_inputs")
x = PositionalEmbedding(sequence_length, vocab_size, embed_dim)
                (decoder_inputs)
x = TransformerDecoder(embed_dim, latent_dim, num_heads)
                (x, encoded_seq_inputs)
x = layers.Dropout(0.5)(x)
decoder_outputs = layers.Dense(vocab_size, activation="softmax")(x)
decoder = keras.Model([decoder_inputs, encoded_seq_inputs],
                decoder_outputs)

decoder_outputs = decoder([decoder_inputs, encoder_outputs])
transformer = keras.Model(
    [encoder_inputs, decoder_inputs], decoder_outputs,
    name="transformer"
)
```

Similar to other deep learning models that have encoder and decoder structures, Transformer models can be used in multiple problem settings. Encoder-only Transformer models can be used for problems like classification. Decoder-only Transformer models can be used for problems like language modeling. Encoder–Decoder Transformer models can be used for problems like machine translation, often with multiple multi-head self-attention modules. These can consist of typical multi-head self-attention blocks which attend to keys and values derived from the same tensor from which queries are learned, found in both the encoder and decoder, but also encoder–decoder cross-attention which can allow the decoder to query values from the encoder.

There are three main variants of Transformer:

- The original Transformer from "Attention Is All You Need" (Encoder & Decoder) (Vaswani *et al.*, 2017),
- BERT (Encoder only) (Devlin *et al.*, 2019),
- GPT (Decoder only) (Brown *et al.*, 2020; Radford and Narasimhan, 2018; Radford *et al.*, 2019).

BERT (Bidirectional Encoder Representations from Transformers) is trained as an Auto-Encoder. It uses Masked Language Model (MLM) to corrupt the input, and the objective of the model is to identify the masked token. It also uses self-attention, where each token in an input sentence looks at the bidirectional context (other tokens on left and right of the considered token).

GPT is trained as an autoregressive model. It is trained with a language modeling objective, where the given sequence of tokens is used to predict the next token (thus looking at only the past or left side context). It also uses Masked Attention to make it into an autoregressive model. A simple definition of the Generative Pretrained Transformer (GPT) model can be done in few hundred lines of code.[2]

From GPT-2 and on, the Transformer models started to be used as multi-task learners, where the learning objective was be modified to be task conditional $P(output|input, task)$. For language models, the output, input, and task are all sequences of natural language. This provided the capability of GPT-2 to operate in zero shot task

---

[2]https://github.com/karpathy/minGPT/blob/master/mingpt/model.py.

transfer tasks, where model "understands" new instructions without being explicitly trained on them. For example, for English to French translation task, the model was given an English sentence followed by the word French and a prompt (:). The model was supposed to understand that it is a translation task and give French counterpart of English sentence.

10.1.2   *Use and application of transformers*

In our discussion so far, we have hinted at the use of Transformer models as yet another, albeit state-of-the-art, tool for representation learning. Drawing upon the word2vec analogy, we focused on the aspect of contextual (self)similarity as a way to build a representation that draws closely related "meanings" to be close in terms of some similarity measure. That measure, most conveniently but not necessarily formulated in terms of a metric, such as Euclidean distance or dot product, had to be specified in the loss function that we used to train our neural network. Once the learning was completed, the representation could be frozen and reused for other purposes. So, although the Transformer was trained in an unsupervised or semi-supervised way toward a particular task like predicting the next sample, or complete a sequence after some elements were omitted via masking, one of the big advantages of the training was in the use of the network as a means for representation of the data that can be later used for other tasks. Speaking of musical tasks of Transformer models, we mention their applications both on the symbolic representation (MIDI Transformers) and on audio data (Waveform and Spectral Transformers) in the coming sections. Indeed, the most common application would be generation of continuation of music materials given some initial seed. Other applications include creating variations through the BERT-like models that do not specifically deal with continuation but can offer completions or alteration to a sequence if some of the elements are discarded from the input, creating melodic, harmonic, or rhythmical variations. One other important point to consider is the use of the representation for downstream tasks, such as adaptation or quick, so-called one-shot of few-shot transfer learning that reuses some of the Transformer layers for a different purpose than in the original training task. Commonly, the reuse is done at the last output layer. An input that is processed through the whole network then is input into a different final layer, often a fully connected classification layer or a multi-layer perceptron (MLP) that is trained on a new task while freezing the Transformer parameters. This technique, not necessarily restricted to use with Transformers, is commonly known as *transfer learning*. A different task is changing the input to the network, or changing both the input and output layers without altering or retraining the middle layers. Such tasks are commonly called reprogramming, as they reuse an existing network for a different task rather then simply refining or adapting it to a different domain. The reprogramming is often done with some adversarial or malign purpose in mind, as a way for an attacker to reprogram models for tasks that violate the code of ethics of the original network developer. Regardless of this unethical use, and since we are not dealing in this book with neural network attacks and defenses, we would like to mention just that the possibility of reprogramming opens up some creative possibilities and understanding of the network representation and task performance accomplishments.

Accordingly, transfer learning and (adversarial) reprogramming are two closely related techniques used for repurposing well-trained neural network models for new tasks. Neural networks, when trained on a large dataset for a particular task, learn features that can be useful across multiple related tasks. Transfer learning aims at exploiting this learned representation for adapting a pretrained neural network for an alternate task. Typically, the last few layers of a neural network are modified to map to a new output space, followed by fine-tuning the network parameters on the dataset of the target task. Such techniques are especially useful when there is a limited amount of training data available for the target task. Reprogramming techniques can be viewed as an efficient training method and can be a superior alternative to transfer learning. Particularly, reprogramming might not require the same large amount of new domain data to fine-tune pre-trained neural networks, including repurposing neural networks across different modalities, such as moving between text and images. This surprising result suggests that there is some universal aspect to the Transformer ability to capture self similarities an repetitions in data that operates across domains and might be a basic trait of artificial (and maybe human) intelligence.

Going back to specifics of Transformer representation, looking at its input and output layers reveals different aspects of data structure that the network is able to extract. The final layer of the Transformer creates an embedding for the whole input sequence. If we think about this as a language model, then the output of the Transformer is effectively an embedding representation for the whole sequence. Summarizing a sentence into a vectors, also know as sentence2vec task, has not been efficiently solved prior to introduction of Transformers. We do need to mention that in the RNN, the last hidden state could be considered as a summarization of the whole input sequence into one vector, also sometimes called a "through vector" as it summarized the idea of the thought of the sentence, which is later used as a seed for generating another similarly meaning sentence when used in a seq2seq fashion, such as in language translation. So, despite the limits of the RNN-thought analysis that required inclusion of a gating memory mechanism and later an attention mechanism, it still helps preserve the idea that Transformer output captures the essence of the data, or for our purposes, "guesses the composer's mind", better than earlier architectures.

Another point worth mentioning is the significance of the first embedding layer that processes the data prior to application of the attention heads and further Transformer layers. As we mentioned earlier, embedding is required when inputting categorical data, such as one-hot vectors representing note combinations, without having a vectorical representation with underlying metric structure. In other words, the purpose of embedding is to use context to find a space with some natural distance measure or metric that places similar meanings in nearby locations. Accordingly, the embedding itself is a process that extracts essential information or features from the input data, sample by sample, rendering it into vectors where self-similarity or attention can be detected. The first embedding layer is also significant for media data, such as images or sounds. Although often some signal processing knowledge is applied to the signal, such as pre-processing it by spectral or MFCC analysis before inputting into the neural network, these man-made features are often not powerful enough to capture the detailed structure of the data. Work on CNNs in the visual domains had shown layers of such networks, such as the first convolutional layer of AlexNet, effectively learn edge detectors and other spatial frequency patterns reminiscent of early filters in the human visual system. Similar inspection of first linear embedding layers in visual Transformer models like Vision Transformer (Dosovitskiy *et al.*, 2021) achieves similar and better filters compared to the CNN first layer. With convolutions without dilation, the receptive field is increased linearly. Using self-attention, the interaction captured between pixels' representations is much more distant, with heads attending to the whole input segment (visual patch) already in the early layers. One can possibly justify the performance gain of Transformer based on the early access to data interactions. Attempts to understand the filtering/feature extraction and interaction aspects of Transformer in audio domains are still in early phases of research, although the empirical quality of results in applying Transformer to various audio and MIDI tasks are suggestive that such architecture are important for music. In the programming exercise of this chapter, we address the two aspects of Transformer — we train a predictive sequence model with and without Transformer and then look at the embedding layer to see how the distribution of tokens changes as a result of using the attention layer.

The surprising efficiency of transformers in many complex machine learning tasks gives an interesting perspective on what types of data and information relations such architectures are managing to extract that other methods haven't. Although deep understanding of the properties of transformers is still an open research question, we note that transformers combine several aspects of modeling in one end-to-end very large and complex architecture. With the huge number of parameters and incredibly long training times, maybe it is not surprising that transformers obtain insights into the data that other models were struggling with. This of course comes at the cost of ease of use and flexibility. While working on zero-shot applications built on top of pre-trained models, we still lack the ability for real-time operations on systems with limited memory and computing power.

Focusing on the relative advantages of the transformer we note that it combines multiple representation layers or attention heads, each one providing direct access and connection across all inputs. Accordingly, transformers overcame difficulties faced by recurrent neural networks (RNNs) that model memory in a sequential manner that decays with time since each step attenuates the previous step by passing the hidden state as the input only to its next neighbor to the right (or to the right and left in bidirectional RNNs).

Moreover, Transformers overcome the locality constraint of CNNs since the Query, Value, and Key operate on the whole input rather than limit scope to a local neighborhood. As such, it is able to handle sequential data and broader contextual data that a convolutional neural network lacks. Lastly, transformer can effectively utilize parallel processing hardware systems such as GPUs and TPUs, which make it an attractive model to those who own massive computational resource (something that common musicians usually lack).

Given these considerations, we take a look at the components of Transformer models in an attempt to provide alternative leaner architectures that might lack the massive joint estimation of relations across all the data elements but may still capture some of the representation advantages and contextual/temporal relations that transformers enjoy. Encoder–decoder aspects of Transformer models are important to understanding the different uses of the architecture. These could be also paralleled to different applications of other generative models, depending on the task they are trained to perform, from generation, to continuation, to accompaniment.

The basic transformer operates as a sequence-to-sequence model. The encoder can be seen as a combined embedding and conditioning operation of a decoder that operates in an auto-regressive manner. The encoder and decoder use different types of attention. The encoder attention is global, i.e., it considers all instances of the input data past and present to build its attention map. The decoder uses an auto-regressive attention as it is trained to generate its output sequentially by combining at every step the encoder output with it own generated past. For instance, a sequence of chords can be input into a decoder and a solo improvisation could be provided at the output of the decoder. The decoder will use a shift of one time step to feed the previous generated melody back into the input of the decoder, in parallel with the chord-derived representation from the encoder. This is reminiscent of the way latent codes from the encoder sequence are combined with the past of the decoder output in seq2seq RNN.

In many cases, the input conditioning is not required, as the transformer learns only the structure of the language it is later asked to generate without an external input. The OpenAI GPT family of models provides an arrangement of the transformer block that is capable of doing language modeling. This model discarded the Transformer encoder and used masked self-attention in the input layer that allows only past event to be considered. As mentioned above, the normal self-attention block allows a position to peek at tokens to its right, while masked self-attention prevents the model from looking into the future.

The generation tasks of Transformers often go beyond generation from a random seed or continuation of a short musical starting sequence. Creative musical tasks can be as varied as inpainting or completion of missing musical segments in partially completed musical sketches, creating variations by omitting segments from the input and letting the transformer find new alternatives, or learning to complete musical materials from partial factorized representations, such as completing melodies when only their rhythm is given and so on. In such cases, instead of using the next token as a prediction, the transformer learns to deal with random omissions of data, trying to recover missing blocks, very much in the spirit of contextual embedding learning done by GloVe (Pennington *et al.*, 2014) or other word2vec models that use context to recover a missing word. Such masking models are applied only to the decoder in the BERT language model.

Similarly to GPT, Magenta's Music Transformer uses a decoder-only transformer to generate music with expressive timing and dynamics. Music modeling is done similar to language modeling by letting the model learn music in an unsupervised way and then have it sample outputs to create novel compositions. The representation fed into the transformer is handled by an embedding block. Such embedding can be learned separately, or be done as part of the overall end-to-end training. In the programming exercise accompanying this chapter, we use a Transformer to demonstrate the embedding aspect of the encoding when self-attention is used for the next token prediction task.

10.1.3 *Kernel view of transformers*

Attention can be interpreted as a means of biasing the allocation of available computational resources toward the most informative components of a signal. This raises the question of what aspect of musical structure is most informative? Apparently, information and attention are closely linked — we pay attention to informative aspects of data and we retrieve information from the environment by paying attention to important aspects of that data. Good representation captures these important aspects, so paying attention and learning representation are ultimately linked as well. An alternative way to think about these problems is as a database lookup problem. Let each feature–data pair be stored in a database as a key–value entry. Information retrieval becomes a problem of finding a relation between an input query and an appropriate database entry by matching the input query to the database key. In other words, if a pair (feature, data) = (key, value) is stored in a database, the key serves as a label for the retrieving the data values for an incoming query. Pairing an input query to database entry is done according to a similarity between the query and the key. This schematic approach to mapping information provides some important insights about the operation of Transformers:

- The problem of representation learning becomes a problem of designing a kernel that matches input data to features/keys.
- Generating data becomes a lookup of entries according to an input query. This can be done by a relatively simple operation such as extracting the top matching key, or by linear weighting of the attention weights.
- Unlike common neural networks, such as CNN or RNN that requires a fixed size input, transformer queries can be designed to operate in a manner that is independent of the database size.
- The same query can result in different answers depending on the contents of the database using multiple keys (multiple attention heads).
- Finding relations between different data types is possible by designing appropriate mapping between their keys and queries.

   Let us carry this analogy one step further. The attention operation in (10.5) "pools" together multiple data points as values. As such, the attention over the database generates a linear combination of values contained in the database in a regression-like fashion. In fact, this contains the special case of database retrieval where all but one weight is zero. The name attention derives from the fact that the pooling operation pays particular attention to the terms for which the weight is significant. In this context, extraction of the attention weight becomes a similarity kernel $\alpha(q, k)$ relating queries to keys.

In particular, considering Euclidean distance, we arrive at what is known as Gaussian kernel:

$$\alpha(q, k_i) = \exp\left(-\frac{\|q - k_i\|^2}{2\sigma^2}\right).$$

For normalized keys and queries, the exponent can be also expressed in terms of a dot product $-\frac{1}{2}\|q - k_i\|^2 = q^T k_i - 1$. Since attention weights need normalizing, this kernel can expressed in terms of a softmax of a dot product, resulting in attention weights similar to (10.4).

Moreover, since the entries to our database operation are input tokens e, the dot-product attention is done by a transformation of the input token through matrices W_q, W_k. These matrices are learned by the transformer so that they suitably transform between the input and output spaces, also adding a third learned matrix W_v to transform the input tokens to desired output values.

The self-attention mechanism is a key defining characteristic of Transformer models. The mechanism can be viewed as a graph-like inductive bias that connects all tokens in a sequence with a relevance-based pooling operation. In a way, the encoder self-attention changes a local representation of the musical data into a global vector of connections or transition probabilities into other instances of that data. Let us rewrite that self-attention equation as a regression:

$$v = f(q) = \sum_i v_i \frac{\alpha(q, k_i)}{\sum_j \alpha(q, k_j)}, \tag{10.6}$$

where $\alpha(q, k_i)$ is an arbitrary kernel, which in the transformer case is commonly an exponent of a dot product between key and query $(e^{q^T k_i})$. This means that the transformer value function performs a regression calculation (pooling) over all values weighted by the attention value that plays the role of a regression kernel between keys and queries. In statistics, kernel regression is a non-parametric technique to estimate the conditional expectation of a random variable, which in our particular case is done using attention as a kernel that computes similarity between keys and queries, with the regression done over values. When values and keys are equal, (10.6) is the same as kernel regression.

PART 4

Generative Models

Chapter 11

Generative Adversarial Networks

11.1 Generative Models

In many applications, we desire to generate new data following the distribution of training data. This task requires learning the underlying distribution $p(\mathbf{x})$ of the training data. In scenarios where label information is relevant, we may also aim to learn the joint distribution $p(\mathbf{x}, y)$. In contrast, a discriminative model focuses solely on modeling $p(y|\mathbf{x})$, such as the image classifier that we have discussed early in Chapter 1.

The early foundations of generative models can be seen as rooted in Statistical Mechanics and Energy-Based Models. Boltzmann Machines (1980s) were developed by Geoffrey Hinton and others, as stochastic neural networks that utilize concepts from statistical mechanics. They model the distribution of data using an energy function, where the system seeks to minimize energy states. They consist of visible and hidden units and use a Gibbs sampling method to learn the distribution of input data. No wonder that the Nobel Prize in Physics that was given to Hinton for foundational discoveries and inventions that enable machine learning with artificial neural networks mentions the Boltzmann Machine.

Boltzmann machines are probabilistic graphical models and the probability distribution of a Boltzmann machine is usually of form $p(\mathbf{x}) = \frac{1}{Z}\exp(-E(\mathbf{x}))$ where E is an energy term. By learning the parameters of a Boltzmann machine, it becomes possible to generate new samples that resemble the training data distribution.

One weakness of Boltzmann machines is their computational complexity, especially when dealing with large datasets. Learning the parameters of a Boltzmann machine requires performing costly Markov Chain Monte Carlo (MCMC) sampling or using approximate inference methods. This makes training Boltzmann machines challenging and computationally expensive, limiting their scalability to complex tasks.

An extension of Boltzmann Machines, called Helmholtz Machines (1990s), introduced an approximate inference method for generative models. They were designed to learn complex data distributions and implemented a two-stage learning process involving a recognition network and a generative network. This model laid groundwork for more complex architectures in the future.

In Chapter 5, we explored variational autoencoders (VAEs), which can be used as generative models. Introduced by Kingma and Welling in 2013, these models can be viewed as a continuation of the Helmholtz machine idea of learning a distribution over latent variables, while allowing for the generation of new data samples by sampling from this distribution.

Indeed, we can get $p(\mathbf{x}, \mathbf{z}) = p(\mathbf{z})p_\theta(\mathbf{x}|\mathbf{z})$ by successively sampling from $p(\mathbf{z})$ and $p_\theta(\mathbf{x}|\mathbf{z})$. However, one limitation of VAEs is that the continuous latent space often leads to the spread of probability masses, resulting in the generation of samples with lower perceptual quality. This phenomenon is particularly notable in image generation tasks, where VAEs may produce samples that appear blurry or lacking in fine details.

The introduction of *generative adversarial networks* (GANs), proposed by Ian Goodfellow and collaborators in 2014, has provided a powerful and promising approach for generating high-quality samples in an efficient manner.

11.2 Basics of GAN

A GAN (Goodfellow *et al.*, 2014) is a generative model that is adversarially trained. It contains a *generator* G, which maps a random noise $\mathbf{z}$ to $G(\mathbf{z})$ in the data space, and a *discriminator* D, which classifies whether an input is real (from training data) or fake (generated). The generator and the discriminator are trained together. On one hand, the discriminator works hard to distinguish fake samples; on the other hand, the generator works hard to fool the discriminator. Once the training is complete, the generator can be used independently to generate new samples by feeding random noise into it. The discriminator, which acted as a critical component during training, is typically discarded or used for other purposes such as evaluation or fine-tuning.

Let's try to understand the intuition behind GANs using the following scenario. Imagine that a mischievous cat and its crafty owner engage in a competitive game. In this game, the cat's ultimate desire is to catch and enjoy delicious fish, while the owner's goal is to tease the cat by drawing fake fish that look realistic but are not actually edible. In this analogy, let's consider the cat as the discriminator and the owner as the generator in the GAN framework.

The game starts with the owner drawing a fish, trying to make it as realistic as possible. This fish represents a generated sample by the generator network. The cat examines the drawn fish and tries to distinguish whether it's a real fish or a fake one. Initially, the cat might easily identify the drawn fish as fake due to discrepancies between the drawing and a real fish. The owner takes note of the cat's feedback and tries to improve their drawing skills to make the fish more convincing. As the training progresses, the owner becomes more skilled at drawing fish, making them visually similar to real fish. The cat, being a keen observer, starts to find it increasingly challenging to distinguish the real fish from the fake ones drawn by the owner. The owner's goal is to continuously improve their drawing skills to fool the cat into thinking that the drawn fish is genuine. Simultaneously, the cat becomes more discerning, honing its ability to identify the subtle differences between real and fake fish. The iterative competition continues until the owner's drawing skills become so refined that the cat can no longer reliably distinguish between real and fake fish. At this point, the cat and the owner have reached a sort of equilibrium. The owner has become an expert at generating realistic-looking fish, while the cat struggles to differentiate them from real fish.

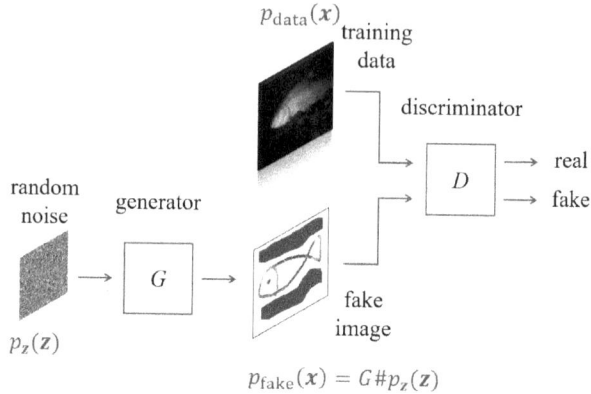

Fig. 11.1. Illustration of a GAN.

Figure 11.1 illustrates how a GAN works.

When training a GAN, we inherit the loss function from the cross-entropy loss of the binary classifier D, where D outputs 1 if the input is real and 0 if the input is fake. Suppose samples come from a mixture of real and fake data with equal weights, up to a constant multiple, such loss is given by $-L(D, G)$, where

$$L(D, G) = \mathbb{E}_{\mathbf{x} \sim p_{\text{data}}(\mathbf{x})}[\log D(\mathbf{x})] + \mathbb{E}_{\mathbf{z} \sim p_{\mathbf{z}}(\mathbf{z})}[\log(1 - D(G(\mathbf{z})))].$$

(11.1)

D and G play the min-max game in which D minimizes $-L(D, G)$ and G maximizes $-L(D, G)$. Equivalently, the optimization problem for GAN is

$$\min_{G} \max_{D} L(D, G). \qquad (11.2)$$

In practice, this can be done by alternatively taking gradient steps for D and G. When taking a step for D, we need to sample the same number of examples from real and generated data. On the other hand, when taking a step for G, we only need to sample from generated data since the gradient with respect to G only has to do with the second term in (11.1). We can choose to associate each gradient step for D with a step for G. Alternatively, we can perform several gradient steps for D before performing a step for G.

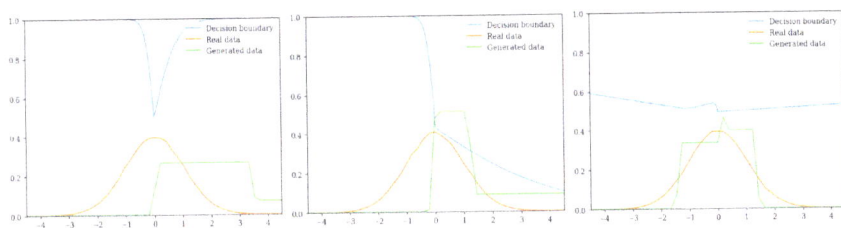

Fig. 11.2. Training a GAN for data from a 1D Gaussian.

Figure 11.2 shows an example where a GAN[1] is trained on data following a 1D Gaussian distribution. In each subplot, the orange curve represents the distribution (histogram) of real data in 1D, while the green curve represents the distribution (histogram) of data in 1D, produced by the generator G of the GAN. That is, in regions where the green curve is higher than the orange curve, it is more likely to see generated data than real data, and vice versa. The aim is to train the GAN so that the green curve approximates the orange curve. Additionally, the blue curve illustrates the output of the discriminator D of the GAN, representing the probability of a sample being real in 1D.

In the left subplot, neither D nor G is well trained. Even in regions where there are more generated data than real data, D may assign a value very close to 1, failing to accurately distinguish between real and generated data. Meanwhile, the differences between the red and green curves are significant. Moving to the middle subplot, D performs well, correctly discerning whether a sample is more likely to be real or generated. However, G is not yet well trained, resulting in notable deviations between the green and orange curves. Finally, in the right subplot, both D and G are well trained. At this stage, D is deceived and unable to differentiate between generated and real data, consistently assigning values close to 0.5. This showcases the training progress of the GAN and the interplay between the discriminator and the generator throughout the learning process.

To have a better understanding of the design in (11.2), let's revisit the goal of generative models, which is to learn the distribution

---

[1]Trained using codes adapted from https://github.com/togheppi/vanilla_GAN and https://github.com/aifromphytsai/gan_sampling.

$p_{\text{data}}(\mathbf{x})$ of the training data. In the context of GANs, we first have a random distribution $p_{\mathbf{z}}(\mathbf{z})$ in a latent space. Then, the generator pushes every sample $\mathbf{z}$ to $G(\mathbf{z})$, which naturally induces a distribution in the data space, which we denote by $p_G := G\sharp p_{\mathbf{z}}$, such that the corresponding probability measures satisfy

$$\mathbb{P}_G(\mathbb{A}) = \mathbb{P}_{\mathbf{z}}(G^{-1}(\mathbb{A})), \quad \text{for all event } \mathbb{A}. \tag{11.3}$$

The question is how close this p_G is to p_{data}. Since p_{data} and p_G are both in the data space, we can first write (11.1) as

$$L(D, G) = \int_{\mathbf{x}} [\log D(\mathbf{x})p_{\text{data}}(\mathbf{x}) + \log(1 - D(\mathbf{x}))p_G(\mathbf{x})]d\mathbf{x}. \tag{11.4}$$

Fix G. For each $\mathbf{x}$, setting the derivative of the integrand with respect to D to zero yields

$$\frac{p_{\text{data}}(\mathbf{x})}{D(\mathbf{x})} = \frac{p_G(\mathbf{x})}{1 - D(\mathbf{x})}. \tag{11.5}$$

The solution gives an optimal D corresponding to each G, which we denote as $D^*(\mathbf{x}; G)$:

$$D^*(\mathbf{x}; G) = \frac{p_{\text{data}}(\mathbf{x})}{p_{\text{data}}(\mathbf{x}) + p_G(\mathbf{x})}. \tag{11.6}$$

Intuitively, D^* implies the likelihood that $\mathbf{x}$ is real given both data distribution and the distribution produced by the generator. It's important to note that D^* depends on the generator G. Now, if we consider the optimization problem in (11.2), we can simplify it further:

$$\min_{G} \max_{D} L(D, G) \tag{11.7}$$

$$= \min_{G} \mathbb{E}_{\mathbf{x} \sim p_{\text{data}}}[\log D^*(\mathbf{x}; G)] + \mathbb{E}_{\mathbf{x} \sim p_G}[\log(1 - D^*(\mathbf{x}; G))] \tag{11.8}$$

$$= \min_{G} \mathbb{E}_{\mathbf{x} \sim p_{\text{data}}}\left[\log \frac{p_{\text{data}}(\mathbf{x})}{p_{\text{data}}(\mathbf{x}) + p_G(\mathbf{x})}\right]$$

$$+ \mathbb{E}_{\mathbf{x} \sim p_G}\left[\log \frac{p_G(\mathbf{x})}{p_{\text{data}}(\mathbf{x}) + p_G(\mathbf{x})}\right]. \tag{11.9}$$

The two expectations in (11.9) are very close to expressions of two KL divergences, which we introduced in Chapter 5. Recall that for two densities p and q,

$$\mathrm{KL}(p\|q) = \mathbb{E}_{\mathbf{x} \sim p} \left[\log \frac{p(\mathbf{x})}{q(\mathbf{x})} \right]. \tag{11.10}$$

Although $p_{\mathrm{data}}(\mathbf{x}) + p_G(\mathbf{x})$ in (11.9) is not a density, we can introduce the average distribution $p_m(\mathbf{x}) = (p_{\mathrm{data}}(\mathbf{x}) + p_G(\mathbf{x}))/2$. By doing so, we can rewrite (11.9) as

$$\min_G \mathrm{KL}(p_{\mathrm{data}}\|p_m) + \mathrm{KL}(p_G\|p_m) - 2\log 2. \tag{11.11}$$

Since the KL divergence is non-negative, and $\mathrm{KL}(p\|q) = 0$ if and only if $p = q$, the minimizer of (11.11) satisfies $p_{\mathrm{data}} = p_G = p_m$. This implies that a GAN is capable of learning the distribution of the training data, as the generated distribution p_G approximates the data distribution p_{data}.

11.3 DCGAN

In this section, we introduce a popular GAN model called Deep Convolutional Generative Adversarial Network, or DCGAN for short, originally introduced by Radford *et al.* (2016). As its name suggests, the generator and discriminator of DCGAN consists of convolutional layers. This architecture has proven to be highly effective in generating realistic images. However, it is important to note that DCGAN is not inherently unique or fundamentally different from other GAN models. Its distinguishing feature lies in the specific choices of architecture and configuration that leverage convolutional layers. To help illustrate DCGAN, especially its architecture, we refer to the PyTorch example provided in their official repository[2] https://github.com/pytorch/examples/tree/main/dcgan. In the original DCGAN code, parallel computation with multiple GPUs is enabled to accelerate training. However, for the purpose of focusing

---

[2]Also, see the tutorial at https://pytorch.org/tutorials/beginner/dcgan_faces_tutorial.html.

on the architecture itself, we omit the details related to multi-GPU parallelization in our discussion.

We first look at the definition of the generator of DCGAN in the code:

```python
class Generator(nn.Module):
    def __init__(self):
        super(Generator, self).__init__()
        self.main = nn.Sequential(
            # input is Z, going into a convolution
            nn.ConvTranspose2d(nz, ngf * 8, 4, 1, 0, bias=False),
            nn.BatchNorm2d(ngf * 8),
            nn.ReLU(True),
            # state size. (ngf*8) x 4 x 4
            nn.ConvTranspose2d(ngf * 8, ngf * 4, 4, 2, 1, bias=False),
            nn.BatchNorm2d(ngf * 4),
            nn.ReLU(True),
            # state size. (ngf*4) x 8 x 8
            nn.ConvTranspose2d(ngf * 4, ngf * 2, 4, 2, 1, bias=False),
            nn.BatchNorm2d(ngf * 2),
            nn.ReLU(True),
            # state size. (ngf*2) x 16 x 16
            nn.ConvTranspose2d(ngf * 2, ngf, 4, 2, 1, bias=False),
            nn.BatchNorm2d(ngf),
            nn.ReLU(True),
            # state size. (ngf) x 32 x 32
            nn.ConvTranspose2d(ngf, nc, 4, 2, 1, bias=False),
            nn.Tanh()
            # state size. (nc) x 64 x 64
        )

    def forward(self, input):
        output = self.main(input)
        return output
```

In this excerpt of the code, `nz` refers to the dimension of the random noise `z` and `ngf` refers to a basis number for the number of channels used in the convolutional layers. Here, `ConvTranspose2d` refers to the two-dimensional *transposed convolution* operation, also known as deconvolution. It performs an "inverse operation" of convolution, taking a low-resolution input and upsampling it to a higher resolution by applying learned filters. The first two arguments in the

`ConvTranspose2d` function refer to the number of input and output channels, respectively; whereas the 3rd to 5th arguments refer to the kernel size, stride steps, and padding, respectively. Each transposed convolution, except for the last layer, is followed by batch normalization and ReLU activation. The output is of size `nc` × 64 × 64, where `nc` = 3 for color images.

Next, we look at the definition of the discriminator of DCGAN:

```python
class Discriminator(nn.Module):
    def __init__(self):
        super(Discriminator, self).__init__()
        self.main = nn.Sequential(
            # input is (nc) x 64 x 64
            nn.Conv2d(nc, ndf, 4, 2, 1, bias=False),
            nn.LeakyReLU(0.2, inplace=True),
            # state size. (ndf) x 32 x 32
            nn.Conv2d(ndf, ndf * 2, 4, 2, 1, bias=False),
            nn.BatchNorm2d(ndf * 2),
            nn.LeakyReLU(0.2, inplace=True),
            # state size. (ndf*2) x 16 x 16
            nn.Conv2d(ndf * 2, ndf * 4, 4, 2, 1, bias=False),
            nn.BatchNorm2d(ndf * 4),
            nn.LeakyReLU(0.2, inplace=True),
            # state size. (ndf*4) x 8 x 8
            nn.Conv2d(ndf * 4, ndf * 8, 4, 2, 1, bias=False),
            nn.BatchNorm2d(ndf * 8),
            nn.LeakyReLU(0.2, inplace=True),
            # state size. (ndf*8) x 4 x 4
            nn.Conv2d(ndf * 8, 1, 4, 1, 0, bias=False),
            nn.Sigmoid()
        )

    def forward(self, input):
        output = self.main(input)
        return output.view(-1, 1).squeeze(1)
```

Here, `ndf` refers to a basis number for the number of channels used in the convolutional layers. The discriminator employs regular convolutional layers, batch normalization and leaky ReLU activation functions. At the end, the output is one dimensional for each sample in the batch.

In the following, excerpt of the code, the loss function, random noise, and the optimizers are specified. The variables `opt.xxx` are specified as hyperparameters by the user:

```
criterion = nn.BCELoss()

fixed_noise = torch.randn(opt.batchSize, nz, 1, 1, device=device)
real_label = 1
fake_label = 0

# setup optimizer
optimizerD = optim.Adam(netD.parameters(), lr=opt.lr,
    betas=(opt.beta1, 0.999))
optimizerG = optim.Adam(netG.parameters(), lr=opt.lr,
    betas=(opt.beta1, 0.999))
```

We see that the binary cross-entropy loss is used. Also, DCGAN employs the Adam optimizer.

Suppose we already defined the generator as `netG` and the discriminator as `netD`. The next excerpt executes the training process:

```
for epoch in range(opt.niter):
    for i, data in enumerate(dataloader, 0):
        ### Update D network
        # train with real data
        netD.zero_grad()
        real_cpu = data[0].to(device)
        batch_size = real_cpu.size(0)
        label = torch.full((batch_size,), real_label,
                    dtype=real_cpu.dtype, device=device) #
                        here, label is 1 for real data,
                        consistent with D minimizing L(D,G)
        output = netD(real_cpu)
        errD_real = criterion(output, label)
        errD_real.backward()
        D_x = output.mean().item()

        # train with generated data
        noise = torch.randn(batch_size, nz, 1, 1, device=device)
        fake = netG(noise)
        label.fill_(fake_label) # here, label is 0 for generated
            data, consistent with D minimizing L(D,G)
        output = netD(fake.detach())
        errD_fake = criterion(output, label)
        errD_fake.backward()
```

```
errD = errD_real + errD_fake
optimizerD.step()

### Update G network
netG.zero_grad()
label.fill_(real_label) # note that generated data are still
    being used here, but we feed label 1, consistent with G
    maximizing L(D,G)
output = netD(fake)
errG = criterion(output, label)
errG.backward()
optimizerG.step()
```

Problem 11.1. According the above box of code excerpt, answer the following questions:

(1) Explain in your own words the roles played by `criterion` in the minimization and maximization optimization of the GAN's loss function, $L(D, G)$.
(2) How many gradient steps are taken for D before a step is taken for G? Can you suggest modifications to the code that would allow for n gradient steps to be taken for D before a step is taken for G?

11.4 f-GAN

According to the above discussion, the loss function of GAN aims to minimize the KL divergence between the data distribution and the generated distribution (more precisely, the average distribution). In general, we can adopt a similar approach and design a GAN by considering a different "distance" between distributions. One such distance measure is the f-divergence (Nowozin *et al.*, 2016), which provides a generalization of the KL divergence.

The f-divergence is determined by a convex function $f : \mathbb{R}_+ \to \mathbb{R}$ such that $f(1) = 0$. Given two probability densities p and q, the f-divergence is defined to be

$$D_f(p\|q) = \mathbb{E}_{\mathbf{x} \sim q}\left[f\left(\frac{p(\mathbf{x})}{q(\mathbf{x})}\right)\right]. \tag{11.12}$$

For example, the KL divergence corresponds to $f(u) = u \log u$. However, there are other choices for f that result in different divergences. Some examples include using $f(u) = -\log u$ to obtain the reverse KL divergence, $f(u) = (u-1)^2$ to form the Pearson chi-squared divergence, or $f(u) = (\sqrt{u}-1)^2$ to form the squared Hellinger divergence. These different choices of f allow us to measure the "distance" between distributions in various ways, each with its own characteristics and applications.

Given a function $f : \mathbb{R}_+ \to \mathbb{R}$ such that $f(1) = 0$, its convex conjugate f^* is defined as

$$f^*(t) = \sup_u ut - f(u). \tag{11.13}$$

If f is a lower semi-continuous function, it is a fact that the biconjugate $f^{**} := (f^*)^*$ is f itself. That is,

$$f(u) = \sup_t tu - f^*(t). \tag{11.14}$$

In this case, we can write

$$D_f(p\|q) = \int q(\mathbf{x}) \sup_t \left(t\frac{p(\mathbf{x})}{q(\mathbf{x})} - f^*(t) \right) d\mathbf{x} \tag{11.15}$$

$$= \int \sup_t (p(\mathbf{x})t - q(\mathbf{x})f^*(t)) d\mathbf{x}. \tag{11.16}$$

In the integrand, a supremum over t is taken for each $\mathbf{x}$. To avoid confusion, we explicitly use $T(\mathbf{x})$ in place of t and write

$$D_f(p\|q) = \int \sup_T (p(\mathbf{x})T(\mathbf{x}) - q(\mathbf{x})f^*(T(\mathbf{x})))d\mathbf{x}. \tag{11.17}$$

Obviously, the right-hand side of (11.17) is no smaller than

$$\int p(\mathbf{x})T(\mathbf{x}) - q(\mathbf{x})f^*(T(\mathbf{x}))d\mathbf{x} \tag{11.18}$$

for any choice of T in (11.18). Therefore,

$$D_f(p\|q) \geq \sup_T \int p(\mathbf{x})T(\mathbf{x}) - q(\mathbf{x})f^*(T(\mathbf{x}))d\mathbf{x} \tag{11.19}$$

$$= \sup_T \mathbb{E}_{\mathbf{x}\sim p}[T(\mathbf{x})] - \mathbb{E}_{\mathbf{x}\sim q}[f^*(T(\mathbf{x}))]. \tag{11.20}$$

We can observe that the form of (11.20) is very similar to (11.1).

Problem 11.2. Explain how you can build a neural network similar to GAN using (11.20).

11.5 Difficulty with GAN

Despite the effectiveness of GANs in generative tasks, there are several challenges associated with them.

One problem is called mode collapse. It refers to a situation where the generator is incapable of producing diverse and varied samples and instead generates limited or repetitive outputs. It occurs when the generator converges to a limited set of modes in the target distribution, neglecting other modes or patterns. Mode collapse in GANs can occur due to poor training dynamics, where the training process gets stuck in a suboptimal equilibrium of the optimization problem. This often happens during the early stages of training when GANs are particularly unstable.

Another problem in GANs is gradient vanishing. When the real data distribution and the generated samples have disjoint supports, it is possible to have a perfect discriminator D^* that assigns a value of 1 to real data and a value of 0 to generated instances. In such cases, the gradient of D^* becomes zero because it has a constant value on both real and fake data. This leads to difficulties in training the generator, as it receives weak or no gradients to guide its updates.

Even when D is not exactly D^*, the gradient may still be too small and thus hinders stable training. Moreover, the gradient of the generator may also be too small. Such a result is formulated by Arjovsky and Bottou (2017). While a detailed mathematical analysis is beyond the scope of this book, we can provide an intuitive understanding.

Let $G_{\boldsymbol{\theta}}$ denote the generator, where $\boldsymbol{\theta}$ represents the trainable parameters in the generator. According to (11.1), the gradient of the generator is $\nabla_{\boldsymbol{\theta}} \mathbb{E}_{\mathbf{z} \sim p_{\mathbf{z}}(\mathbf{z})}[\log(1 - D(G_{\boldsymbol{\theta}}(\mathbf{z})))]$. We can safely switch $\nabla_{\boldsymbol{\theta}}$ and $\mathbb{E}$ and apply the chain rule to get

$$\nabla_{\boldsymbol{\theta}} \mathbb{E}_{\mathbf{z} \sim p_{\mathbf{z}}(\mathbf{z})}[\log(1 - D(G_{\boldsymbol{\theta}}(\mathbf{z})))] \tag{11.21}$$

$$= \mathbb{E}_{\mathbf{z} \sim p_{\mathbf{z}}(\mathbf{z})}\left[\frac{-\nabla_{\boldsymbol{\theta}} D(G_{\boldsymbol{\theta}}(\mathbf{z}))}{1 - D(G_{\boldsymbol{\theta}}(\mathbf{z}))}\right] \tag{11.22}$$

$$= -\mathbb{E}_{\mathbf{z} \sim p_{\mathbf{z}}(\mathbf{z})} \left[\frac{\mathbf{J}_{\boldsymbol{\theta}}^{\top}(G_{\boldsymbol{\theta}}(\mathbf{z})) \nabla D(G_{\boldsymbol{\theta}}(\mathbf{z}))}{1 - D(G_{\boldsymbol{\theta}}(\mathbf{z}))} \right], \tag{11.23}$$

where $\mathbf{J}_{\boldsymbol{\theta}}(G_{\boldsymbol{\theta}}(\mathbf{z}))$ is the Jacobian matrix of G.

Suppose D is close to D^* in the sense that both $\|\nabla D - \nabla D^*\| \le \epsilon$ and $|D - D^*| \le \epsilon$ hold (uniformly) for some small $\epsilon > 0$. Continuing with (11.23), we have

$$\|\nabla_{\boldsymbol{\theta}} \mathbb{E}_{\mathbf{z} \sim p_{\mathbf{z}}(\mathbf{z})} [\log(1 - D(G_{\boldsymbol{\theta}}(\mathbf{z})))]\|^2 \tag{11.24}$$

$$\le \mathbb{E}_{\mathbf{z} \sim p_{\mathbf{z}}(\mathbf{z})} \left[\frac{\|\mathbf{J}_{\boldsymbol{\theta}}(g_{\boldsymbol{\theta}}(\mathbf{z}))\|^2 (\|\nabla D^*(G_{\boldsymbol{\theta}}(\mathbf{z}))\| + \epsilon)^2}{(|1 - D^*(G_{\boldsymbol{\theta}}(\mathbf{z}))| - \epsilon)^2} \right] \tag{11.25}$$

$$= \mathbb{E}_{\mathbf{z} \sim p_{\mathbf{z}}(\mathbf{z})} \left[\frac{\|\mathbf{J}_{\boldsymbol{\theta}}(g_{\boldsymbol{\theta}}(\mathbf{z}))\|^2 \epsilon^2}{(1 - \epsilon)^2} \right] \tag{11.26}$$

$$= \left(\frac{\epsilon}{1 - \epsilon} \right)^2 \mathbb{E}_{\mathbf{z} \sim p_{\mathbf{z}}(\mathbf{z})} [\|\mathbf{J}_{\boldsymbol{\theta}}(g_{\boldsymbol{\theta}}(\mathbf{z}))\|^2], \tag{11.27}$$

where Jensen's inequality has been applied so that we can switch the expectation and the squared norm in (11.25).

Effective training of GANs has been a subject of extensive research, and various techniques and "tricks" have been proposed to improve training stability and quality of generated samples. For instance, the DCGAN paper and subsequent works have provided valuable insights and recommendations for training GANs effectively.

As seen from the above discussion, one common observation is that the logarithmic function used in the original GAN loss function can introduce challenges during training. To address this issue, alternative loss functions have been proposed that alleviate the problems associated with the logarithm. By adopting different loss functions, researchers have aimed to overcome the challenges associated with the original GAN formulation and improve the training process. These alternative loss functions often provide more stable and reliable gradients, allowing for better convergence and reducing the likelihood of mode collapse.

In the following chapter, we explore one of these alternative loss functions in detail, namely the Wasserstein GAN (WGAN).

Chapter 12

Wasserstein Generative Adversarial Network

12.1 Distance between Distributions

In Chapter 11, we examined some of the challenges associated with GANs. One notable issue we discussed was gradient vanishing, which arises from the implicit minimization of the divergence between the real and fake distributions in GANs.

Can we consider an alternative approach to defining the "distance between distributions"? Let's analogize abstract distributions to distributions of commodities. Consider a scenario where we have a large number of commodities distributed around Boston and we want to transport them to different locations, such as San Francisco and Maine. It is evident that the cost of transportation between Boston and San Francisco will be much higher than between Boston and Maine. This disparity in transportation costs suggests that the notion of distance between distributions should account for these varying costs.

In (11.3), we introduced the concept of the "push forward" operation. Suppose that we have two distributions p and q over $\mathbb{R}^n$ where $q = T\sharp p$ is pushed forward from p by a transformation $T : \mathbb{R}^n \to \mathbb{R}^n$. The important idea that will finally resolve the problem with GANs is that the transformation is not "free". Moving a point $\mathbf{x} \in \mathbb{R}^n$ to $T(\mathbf{x}) \in \mathbb{R}^n$ incurs a cost $c(\mathbf{x}, T(\mathbf{x}))$. Since we need to consider all the points in $\mathbb{R}^n$, the total cost of pushing p toward q is

$$\int c(\mathbf{x}, T(\mathbf{x}))p(\mathbf{x})d\mathbf{x}. \tag{12.1}$$

To define the distance between p and q, we can consider the minimum total cost of pushing p toward q. That is, we need to solve the following optimization problem:

$$\min_{T} \int c(\mathbf{x}, T(\mathbf{x}))p(\mathbf{x})d\mathbf{x} \tag{12.2}$$

$$\text{s.t. } q = T\sharp p. \tag{12.3}$$

This optimization problem is known as the Monge problem, which has a long history in optimal transport theory. It dates back to the 18th century and is named after the French mathematician Gaspard Monge. Monge first formulated the problem while studying optimal ways to transport soil during civil engineering projects. The Monge problem seeks to find an optimal mapping between two probability measures that minimizes the total cost of transportation.

However, the Monge problem has a limitation in that the function T can only assign commodities to a single destination. In our example, this means that we cannot transport commodities from Boston (represented by $\mathbf{x}$) to both San Francisco and Maine simultaneously using a single $T(\mathbf{x})$ function.

To this end, the Monge problem was extended by the Russian mathematician Leonid Kantorovich in the 20th century. He introduced the notion of a transportation plan, which relaxes the Monge problem. Specifically, we consider the cost $c(\mathbf{x}, \mathbf{y})$ for every possible pair of $\mathbf{x}$ and $\mathbf{y}$. The objective is to find an optimal transportation plan that minimizes the overall cost. For a pair of $\mathbf{x}$ and $\mathbf{y}$, let $\pi(\mathbf{x}, \mathbf{y})$ denote the joint distribution, that is, the *transportation plan*. To ensure the validity of the transportation plan, we require p and q be the marginal distributions of π, that is,

$$\int \pi(\mathbf{x}, \mathbf{y})d\mathbf{y} = p(\mathbf{x}) \quad \text{and} \quad \int \pi(\mathbf{x}, \mathbf{y})d\mathbf{x} = q(\mathbf{y}). \tag{12.4}$$

Similar to the Monge problem, the distance between the distributions p and q can be derived by solving an optimization problem. The objective is to minimize the integral of the cost function $c(\mathbf{x}, \mathbf{y})$ weighted by the transportation plan $\pi(\mathbf{x}, \mathbf{y})$. This can be expressed as follows:

$$\min_{\pi} \int c(\mathbf{x}, \mathbf{y}) \pi(\mathbf{x}, \mathbf{y}) d\mathbf{x} d\mathbf{y} \qquad (12.5)$$

$$\text{s.t.} \int \pi(\mathbf{x}, \mathbf{y}) d\mathbf{y} = p(\mathbf{x}), \qquad (12.6)$$

$$\int \pi(\mathbf{x}, \mathbf{y}) d\mathbf{x} = q(\mathbf{y}). \qquad (12.7)$$

This is the famous Kantorovich problem. We use $\Pi(p, q)$ to denote the set of transportation plans π that satisfies (12.6) and (12.7). When taking $c(\mathbf{x}, \mathbf{y}) = \|\mathbf{x} - \mathbf{y}\|^{\nu}$, where $\nu \geq 1$, the optimal value of (12.5) defines the ν-*Wasserstein distance* between p and q. More specifically, the ν-Wasserstein distance between p and q is defined to be

$$W_{\nu}(p, q) := \left\{ \min_{\pi \in \Pi(p,q)} \int \|\mathbf{x} - \mathbf{y}\|^{\nu} \pi(\mathbf{x}, \mathbf{y}) d\mathbf{x} d\mathbf{y} \right\}^{1/\nu}. \qquad (12.8)$$

When $\nu = 1$, W_1 is also known as the "earth-mover" distance (EMD):

$$\text{EMD}(p, q) := W_1(p, q) = \min_{\pi \in \Pi(p,q)} \int \|\mathbf{x} - \mathbf{y}\| \pi(\mathbf{x}, \mathbf{y}) d\mathbf{x} d\mathbf{y}. \qquad (12.9)$$

The Wasserstein distance is widely used in logistics, supply chain management, economics, and machine learning.

12.2 Deriving the Loss Function*

Suppose we have chosen the Wasserstein distance W_{ν} as our metric for generative modeling. In this context, our goal is to learn a data distribution p by generating a distribution q that minimizes the Wasserstein distance $W_{\nu}(p, q)$ between them. However, it may not be immediately clear how to effectively utilize the formulation of the Wasserstein distance as given in (12.8).

The key technique utilized in this context is known as the *Kantorovich–Rubinstein duality* for $\nu = 1$, which states that

$$W_1(p, q) = \sup_{\|f\|_{\text{Lip}} \leq 1} \mathbb{E}_{\mathbf{x} \sim p} f(\mathbf{x}) - \mathbb{E}_{\mathbf{y} \sim q} f(\mathbf{y}), \qquad (12.10)$$

where

$$\|f\|_{\text{Lip}} := \sup_{\mathbf{x} \neq \mathbf{y}} \frac{|f(\mathbf{x}) - f(\mathbf{y})|}{\|\mathbf{x} - \mathbf{y}\|} \qquad (12.11)$$

is the Lipschitz norm of f.

At first glance, it may not be immediately apparent that (12.10) corresponds to (12.9). Establishing their rigorous equivalence requires a detailed proof. However, we can provide an overview of the main idea behind the proof to illustrate their connection.

We first introduce the following more general *Kantorovich duality*: Define $I(\pi)$ by

$$I(\pi) = \int c(\mathbf{x}, \mathbf{y}) \pi(\mathbf{x}, \mathbf{y}) d\mathbf{x} d\mathbf{y}, \qquad (12.12)$$

and define $J(\phi, \psi)$ by

$$J(\phi, \psi) = \mathbb{E}_{\mathbf{x} \sim p} \phi(\mathbf{x}) + \mathbb{E}_{\mathbf{y} \sim q} \psi(\mathbf{y}) = \int \phi(\mathbf{x}) p(\mathbf{x}) d\mathbf{x} + \int \psi(\mathbf{y}) q(\mathbf{y}) d\mathbf{y}. \qquad (12.13)$$

We impose the condition that $\pi \in \Pi(p, q)$, while for ϕ and ψ, we require that $(\phi, \psi) \in \Phi_c$, where $\Phi_c = \{(\phi, \psi) : \phi(\mathbf{x}) + \psi(\mathbf{y}) \leq c(\mathbf{x}, \mathbf{y}) \text{ for all } \mathbf{x}, \mathbf{y}\}$. The Kantorovich duality refers to the fact that

$$\inf_{\pi \in \Pi(p,q)} I(\pi) = \sup_{(\phi, \psi) \in \Phi_c} J(\phi, \psi). \qquad (12.14)$$

On one hand, it is straightforward to show that $\sup_{(\phi, \psi) \in \Phi_c} J(\phi, \psi) \leq \inf_{\pi \in \Pi(p,q)} I(\pi)$. Indeed, for any $(\phi, \psi) \in \Phi_c$ and any $\pi \in \Pi(p, q)$, we can replace p and q in (12.13) with (12.6) and (12.7) respectively. This allows us to infer that

$$J(\phi, \psi) = \int \phi(\mathbf{x}) \pi(\mathbf{x}, \mathbf{y}) d\mathbf{x} d\mathbf{y} + \int \psi(\mathbf{y}) \pi(\mathbf{x}, \mathbf{y}) d\mathbf{x} d\mathbf{y} \qquad (12.15)$$

$$= \int [\phi(\mathbf{x}) + \psi(\mathbf{y})] \pi(\mathbf{x}, \mathbf{y}) d\mathbf{x} d\mathbf{y} \qquad (12.16)$$

$$\leq \int c(\mathbf{x},\mathbf{y})\pi(\mathbf{x},\mathbf{y})d\mathbf{x}d\mathbf{y} \tag{12.17}$$

$$= I(\pi). \tag{12.18}$$

On the other hand, one can express the problem of minimizing $I(\pi)$ over $\pi \in \Pi(p,q)$ as follows:

$$\inf_{\pi \in \Pi(p,q)} I(\pi) \tag{12.19}$$

$$= \inf_{\pi \geq 0} \int c(\mathbf{x},\mathbf{y})\pi(\mathbf{x},\mathbf{y})d\mathbf{x}d\mathbf{y} \tag{12.20}$$

$$+ \sup_{\phi,\psi} \int \phi(\mathbf{x})p(\mathbf{x})dx + \int \psi(\mathbf{y})q(\mathbf{y})dy$$

$$- \int [\phi(\mathbf{x}) + \psi(\mathbf{y})]\pi(\mathbf{x},\mathbf{y})d\mathbf{x}d\mathbf{y}. \tag{12.21}$$

In (12.21), we replace the condition $\pi \in \Pi(p,q)$ with the terms inside the supremum. It is evident that the supremum in (12.21) is equal to 0 if $\pi \in \Pi(p,q)$ but becomes ∞ otherwise since one can choose the magnitudes of ϕ and ψ as large as possible.

The trick here is to swap the order of "inf" and "sup". This would take some nontrivial work and in general one cannot turn a "min-max" directly into a "max-min". For a rigorous proof, the reader is referred to the work of Villani (2003). We circumvent the technical details here and assume it is legal to first take $\inf_{\pi \geq 0}$ and then $\sup_{\phi,\psi}$. With this rearrangement, we can proceed as follows:

$$\inf_{\pi \in \Pi(p,q)} I(\pi) \tag{12.22}$$

$$= \sup_{\phi,\psi} \int \phi(\mathbf{x})p(\mathbf{x})dx + \int \psi(\mathbf{y})q(\mathbf{y})dy \tag{12.23}$$

$$- \sup_{\pi \geq 0} \left(\int [\phi(\mathbf{x}) + \psi(\mathbf{y})]\pi(\mathbf{x},\mathbf{y})d\mathbf{x}d\mathbf{y} - \int c(\mathbf{x},\mathbf{y})\pi(\mathbf{x},\mathbf{y})d\mathbf{x}d\mathbf{y} \right)$$

$$\tag{12.24}$$

$$= \sup_{\phi,\psi} \int \phi(\mathbf{x})p(\mathbf{x})dx + \int \psi(\mathbf{y})q(\mathbf{y})dy \tag{12.25}$$

$$- \sup_{\pi \geq 0} \int [\phi(\mathbf{x}) + \psi(\mathbf{y}) - c(\mathbf{x}, \mathbf{y})] \pi(\mathbf{x}, \mathbf{y}) d\mathbf{x} d\mathbf{y}. \tag{12.26}$$

If $(\phi, \psi) \in \Phi_c$, then (12.26) is equal to 0; otherwise, (12.26) becomes ∞ since one can assign large value of $\pi(\mathbf{x}, \mathbf{y})$ at $(\mathbf{x}, \mathbf{y})$, where $\phi(\mathbf{x}) + \psi(\mathbf{y}) - c(\mathbf{x}, \mathbf{y})$ is positive. Therefore,

$$\inf_{\pi \in \Pi(p,q)} I(\pi) \tag{12.27}$$

$$= \sup_{(\phi, \psi) \in \Phi_c} \int \phi(\mathbf{x}) p(\mathbf{x}) d\mathbf{x} + \int \psi(\mathbf{y}) q(\mathbf{y}) d\mathbf{y} \tag{12.28}$$

$$= \sup_{(\phi, \psi) \in \Phi_c} J(\phi, \psi). \tag{12.29}$$

In view of the above Kantorovich duality, one can rewrite the W_1 distance as follows:

$$W_1(p, q) = \sup_{(\phi, \psi) \in \Phi_c} J(\phi, \psi), \tag{12.30}$$

where Φ_c is specified to be $\{(\phi, \psi) : \phi(\mathbf{x}) + \psi(\mathbf{y}) \leq \|\mathbf{x} - \mathbf{y}\|$ for all $\mathbf{x}, \mathbf{y}\}$. Note that the form of $J(\phi, \psi) = \mathbb{E}_{\mathbf{x} \sim p} \phi(\mathbf{x}) + \mathbb{E}_{\mathbf{y} \sim q} \psi(\mathbf{y})$ is already very similar to the right-hand side of (12.10). The question is where the Lipschitz norm comes from. To see this, define the "c-transform"

$$\phi^c(\mathbf{y}) = \inf_{\mathbf{x}} [c(\mathbf{x}, \mathbf{y}) - \phi(\mathbf{x})]. \tag{12.31}$$

Applying c-transform again to ϕ^c yields

$$\phi^{cc}(\mathbf{x}) = \inf_{\mathbf{y}} [c(\mathbf{x}, \mathbf{y}) - \phi^c(\mathbf{y})], \tag{12.32}$$

Clearly,

$$\sup_{(\phi, \psi) \in \Phi_c} J(\phi, \psi) = \sup_{\phi} J(\phi^{cc}, \phi^c). \tag{12.33}$$

On the other hand, $\|\phi^c\|_{\text{Lip}} = 1$ when we specify $c(\mathbf{x}, \mathbf{y}) = \|\mathbf{x} - \mathbf{y}\|$ (left as a simple exercise). This allows us to infer:

$$-\phi^c(\mathbf{x}) \leq \inf_{\mathbf{y}} (\|\mathbf{x} - \mathbf{y}\| - \phi^c(\mathbf{y})) \leq \|\mathbf{x} - \mathbf{x}\| - \phi^c(\mathbf{x}) = -\phi^c(\mathbf{x}). \tag{12.34}$$

Since the leftmost and rightmost terms are the same, the "$\leq$" in (12.34) are all "=". Also note the second term in (12.34) is exactly

ϕ^{cc}. We thus have $\phi^{cc} = -\phi^c$. Therefore,

$$\sup_{(\phi,\psi)\in\Phi_c} J(\phi,\psi) = \sup_{\phi} J(\phi^{cc},\phi^c) = \sup_{\phi} J(-\phi^c,\phi^c) \qquad (12.35)$$

$$\leq \sup_{\|f\|_{\mathrm{Lip}}\leq 1} J(f,-f) \leq \sup_{(\phi,\psi)\in\Phi_c} J(\phi,\psi), \qquad (12.36)$$

where again, all "$\leq$" signs become "$=$". Finally, we conclude that

$$W_1(p,q) = \sup_{\|f\|_{\mathrm{Lip}}\leq 1} J(f,-f) = \sup_{\|f\|_{\mathrm{Lip}}\leq 1} \mathbb{E}_{\mathbf{x}\sim p}f(\mathbf{x}) - \mathbb{E}_{\mathbf{y}\sim q}f(\mathbf{y}).$$
$$(12.37)$$

This establishes (12.10).

12.3 Wasserstein GAN with Weight Clipping

By virtue of the Kantorovich-Rubinstein duality (12.10), we can derive an alternative optimization problem for GAN, known as Wasserstein GAN, or WGAN for short (Arjovsky *et al.*, 2017). The objective of WGAN is formulated as follows:

$$\min_{G} \max_{D:\|D\|_{\mathrm{Lip}}\leq 1} L_{W_1}(D,G), \qquad (12.38)$$

where

$$L_{W_1}(D,G) = \mathbb{E}_{\mathbf{x}\sim p_{\mathrm{data}}(\mathbf{x})}D(\mathbf{x}) - \mathbb{E}_{\mathbf{z}\sim p_{\mathbf{z}}(\mathbf{z})}D(G(\mathbf{z})). \qquad (12.39)$$

This formulation introduces two key differences between WGAN and the original GAN framework. First, the loss function (12.39) does not involve logarithmic functions, as seen in (11.1). Second, an additional constraint is imposed on the discriminator, requiring its Lipschitz norm to be bounded by 1, i.e., $\|D\|_{\mathrm{Lip}} \leq 1$. The discriminator D in WGAN is often referred to as the critic because D does not play the role of a binary classifier.

In WGAN, the specific value of the Lipschitz constant in the constraint $\|D\|_{\mathrm{Lip}} \leq 1$ is not essential. We can replace the value 1 with any other fixed positive constant. Indeed, if (D^*, G^*) is an optimal discriminator-generator pair for the optimization problem (12.38), then (KD^*, G^*) would be optimal for the optimization problem should we replace $\|D\|_{\mathrm{Lip}} \leq 1$ in (12.38) with $\|D\|_{\mathrm{Lip}} \leq K$.

Therefore, we will learn the same generator as long as we have a fixed Lipschitz constant of D, which we don't even need to know!

One way to enforce a fixed Lipschitz constant of D is by constraining the weights of the network to a specific range. Parameter clipping is a commonly used technique to achieve this constraint. Specifically, after each gradient update step, we can iterate through the parameters of D and clamp their values within the desired range, say $[-c, c]$. For instance, in PyTorch, this can be easily achieved by calling `torch.clamp` or `torch.Tensor.clamp\_` (the inplace version). Specifically, after each gradient step, one can call

```
for par in D.parameters():
    par.data.clamp_(-c, c)
```

Problem 12.1. Revise `criterion` in the code script is Section 11.3 so that it corresponds to the loss of WGAN. Moreover, incorporate the above code snippet to enable weight clipping.

12.4   Wasserstein GAN with Gradient Penalty

Weight clipping, while simple to implement, can have some drawbacks in terms of model expressivity and potential issues such as mode collapse. In fact, many parameters of the critic will have weight value equal to either c or $-c$ if we implement weight clipping with range $[-c, c]$. This can lead to a significant reduction in the range of weight values, effectively limiting the model's capacity to represent complex functions and capture fine-grained details in the data distribution.

Furthermore, weight clipping can introduce optimization challenges. The constraint imposed by weight clipping is non-smooth and discontinuous, which can make the optimization process more difficult. It may lead to instability, gradient vanishing or exploding, and slower convergence.

To address these issues, several alternative methods have been proposed to enforce the Lipschitz constraint more effectively. One popular approach is to use gradient penalty, as introduced in the WGAN-GP variant (Gulrajani *et al.*, 2017). Instead of directly constraining the weights, the Lipschitz condition is enforced by adding a gradient penalty term to the loss function. This penalty encourages the gradients of the critic with respect to its inputs to have a L2 norm close to 1. By doing so, it indirectly enforces the Lipschitz constraint without resorting to weight clipping.

Problem 12.2. Suppose $f : \mathbb{R}^d \to \mathbb{R}$ is a continuously differentiable function that satisfies $\|f\|_{\mathrm{Lip}} \leq 1$. Is it true that for any $\mathbf{x} \in \mathbb{R}^d$

$$\|\nabla f(\mathbf{x})\|_2 \leq 1? \tag{12.40}$$

The specific optimization problem for WGAN-GP is as follows:

$$\min_G \max_D L_{\mathrm{GP}}(D, G) = \mathbb{E}_{\mathbf{x} \sim p_{\mathrm{data}}(\mathbf{x})} D(\mathbf{x}) - \mathbb{E}_{\mathbf{z} \sim p_{\mathbf{z}}(\mathbf{z})} D(G(\mathbf{z}))$$

$$- \lambda \mathbb{E}_{\hat{\mathbf{x}} \sim p_{\mathrm{interp}}(\hat{\mathbf{x}})} [(\|\nabla_{\hat{\mathbf{x}}} D(\hat{\mathbf{x}})\|_2 - 1)^2]. \tag{12.41}$$

The gradient penalty is imposed as the regularization term. Note that the loss in the formulation above is maximized with respect to the critic D, as indicated by the minus sign in front of the gradient penalty term. Here, $\hat{\mathbf{x}} \sim p_{\mathrm{interp}}(\hat{\mathbf{x}})$ refers to that $\hat{\mathbf{x}}$ is sampled uniformly from the line segment connecting a generated example and an example from the training data. Arjovsky *et al.* (2017) showed that if D^* is an optimal solution of (12.38), then under mild condition, the norm of $\nabla_{\hat{\mathbf{x}}} D^*(\hat{\mathbf{x}})$ is equal to 1 with probability 1 under the transportation plan.

We end this chapter by illustrating the following code snippet for WGAN-GP. We only show the parts related to the gradient penalty:

```python
# ......
# assume netG, netD, optimizerG, optimizerD have been defined
# define the Wasserstein loss
def compute_wasserstein_loss(real_samples, fake_samples):
    real_scores = netD(real_samples)
    fake_scores = netD(fake_samples)
    return torch.mean(fake_scores) - torch.mean(real_scores)


# define the gradient penalty
def compute_gradient_penalty(real_samples, fake_samples):
    alpha = torch.rand(real_samples.size(0), 1, 1, 1).to(device)
    interpolates = (alpha * real_samples + (1 - alpha) *
        fake_samples).requires_grad_(True) # uniformly sample to
        get hat{x}
    gradients = autograd.grad(outputs=netD(interpolates),
```

```
                    inputs=interpolates,
                    grad_outputs=torch.ones_like(netD(interpolates)),
                    create_graph=True,
                    retain_graph=True,
                    only_inputs=True)[0]
    gradient_penalty = ((gradients.norm(2, dim=1) - 1) ** 2).mean()
    return gradient_penalty

# training loop
for epoch in range(opt.niter):
    for i, data in enumerate(dataloader, 0):
        # generate fake samples using the generator
        noise = torch.randn(opt.batchSize, nz, 1, 1, device=device)
        fake = netG(noise)

        # compute the Wasserstein loss
        wasserstein_loss = compute_wasserstein_loss(data, fake)

        # compute the gradient penalty
        gradient_penalty = compute_gradient_penalty(data, fake)

        # compute the total loss
        total_loss = wasserstein_loss + lambda_gp * gradient_penalty
            # lambda_gp is a hyperparameter

        # update the discriminator
        optimizerD.zero_grad()
        total_loss.backward(retain_graph=True)
        optimizerD.step()

        # continue to update the generator, etc.
        # ......
```

Chapter 13

Normalizing Flows and Diffusion Models

13.1 Normalizing Flows

In previous chapters, we have mainly explored two prominent classes of generative models, namely GANs and VAEs. GANs operate with an implicit density representation, meaning that there is no direct description of the underlying density function. Through an adversarial training process, GANs learn to generate realistic samples that capture the distribution characteristics of the training data. In contrast, VAEs take an explicit modeling approach. VAEs utilize the framework of variational inference and employ an encoder–decoder architecture. They employ the ELBO approximation to estimate the intractable posterior distribution.

There are also generative models that aim to model tractable densities. One notable example is the pixelRNN model proposed by Van Den Oord *et al.* (2016). In pixelRNN, our goal is to model the density $p(\mathbf{x})$ of an n-by-n pixel image, where $\mathbf{x} \in \mathbb{R}^{n^2}$ is the random vector that describes all the pixel values. Clearly, $p(\mathbf{x}) = \prod_{i=1}^{n^2} p(x_i|x_1, \ldots, x_{i-1})$. By writing $p(\mathbf{x})$ in this manner, we are imposing a causal relation among the pixels and generation will be done in a sequential manner: x_i will be generated only after $x_1, \ldots, x_{i-1}$ have been generated. Given the sequential manner of the generation process, we can model the density using an RNN and train it according to maximum likelihood. Building upon the pixelRNN model, improved variations have been introduced, such as the pixelCNN. In pixelCNN, additional masking techniques are employed to enforce the causal relation between pixels during the generation process. These masks ensure that each pixel is generated based only on the previously generated pixels, avoiding the use of future pixels that have not been generated yet.

A more general and expressive approach to defining densities is through *normalizing flows*. In a normalizing flow, a chain of simple invertible transformations are cascaded to map the data distribution to a latent and normalized variable. The term "normalizing" refers to the fact that the transformed variable has a total probability of 1, whereas the term "flow" refers to the fact that a chain of simple transformations are used to ensure expressivity. An illustration of the framework is in Figure 13.1. Here, the transformation T^{-1} is responsible for mapping an input $\mathbf{x}$ to a latent variable $\mathbf{z}$ of the same dimension (we use the inverse sign here so that the notation is consistent with the subsequent technical discussion). The inverse of the flow, T, performs the transformation from $\mathbf{z}$ to a reconstructed output $\tilde{\mathbf{x}}$ that ideally recovers the original input $\mathbf{x}$. The transformation T can be constructed using a chain of individual transformations, represented as $T = T_k \circ \cdots \circ T_2 \circ T_1$ in the figure, which implies that $T^{-1} = T_1^{-1} \circ T_2^{-1} \circ \cdots \circ T_k^{-1}$.

Figure 13.2 presents a more illustrative example[1] of a normalizing flow. In this particular case, our goal is to learn the distribution of $\mathbf{x}$, which follows a "double moon" shape. We can interpret $\mathbf{x}$ as a transformed version of a latent variable $\mathbf{z}$ drawn from a Gaussian distribution, where the transformation is denoted as $\mathbf{x} = T(\mathbf{z})$. The function T represents an invertible mapping from the latent space to the observed data space. Ideally, with successful training, the learned transformation T would accurately capture the intricate shape of the

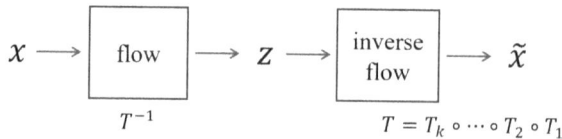

$$x \longrightarrow \boxed{\text{flow}} \longrightarrow z \longrightarrow \boxed{\begin{array}{c}\text{inverse}\\\text{flow}\end{array}} \longrightarrow \tilde{x}$$

$$T^{-1} \qquad\qquad\qquad T = T_k \circ \cdots \circ T_2 \circ T_1$$

Fig. 13.1. Illustration of normalizing flow.

---

[1]The figures are generated using codes adapted from https://github.com/senya-ashukha/real-nvp-pytorch.

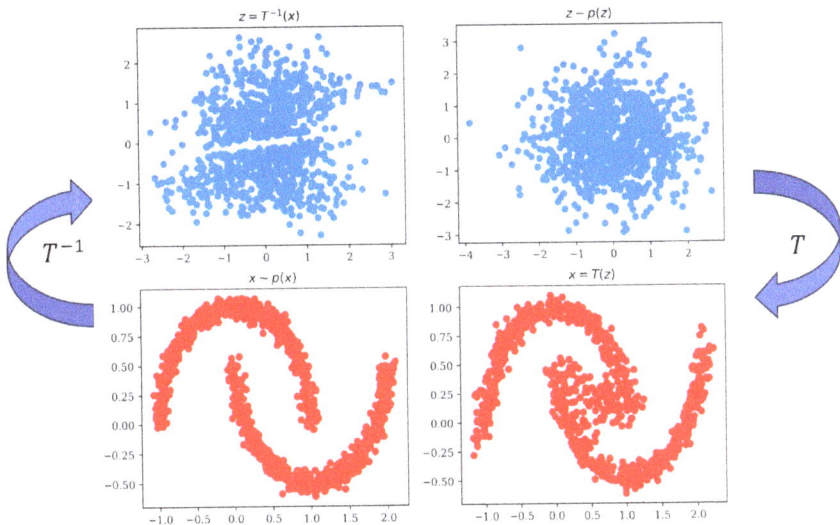

Fig. 13.2. An example of normalizing flow.

"double moon" distribution. However, in the depicted example, the training process is not perfect, leading to an imperfect recovery of the true distribution. As a result, the generated samples $T(\mathbf{z})$ may not precisely match the desired "double moon" shape.

Normalizing flows have attracted significant attention from the machine learning community since around 2015. See the work of Kobyzev *et al.* (2020) and Papamakarios *et al.* (2021) if you are interested in a comprehensive understanding of normalizing flows and their applications. These reviews provide detailed insights into the theoretical foundations, methodologies, and recent advancements in the field of normalizing flows.

Let's first consider a simple block of an invertible transformation, say $\mathbf{x} = T(\mathbf{z})$, where $p_{\mathbf{z}}(\mathbf{z})$ is called the base distribution. Here, T is not only required to be invertible, but both T and T^{-1} need to be differentiable (such T is called a diffeomorphism). In practice, both the transformation T and the base distribution $p_{\mathbf{z}}$ may be parameterized, say $T = T(\cdot\,;\boldsymbol{\phi})$ and $p_{\mathbf{z}} = p_{\mathbf{z}}(\cdot\,;\boldsymbol{\psi})$.

Since $p_{\mathbf{x}} = T\#p_{\mathbf{z}}$, we have $p_{\mathbf{x}}(\mathbf{x})d\mathbf{x} = p_{\mathbf{z}}(\mathbf{z})d\mathbf{z}$ (recall the discussion from Chapter 12). Applying the change of variable $d\mathbf{x} = |\det J_T(\mathbf{z})|\, d\mathbf{z}$, we have

$$p_{\mathbf{x}}(\mathbf{x})\,|\det J_T(\mathbf{z})| = p_{\mathbf{z}}(\mathbf{z}). \tag{13.1}$$

Since T is invertible, following the same reasoning, we have

$$p_{\mathbf{x}}(\mathbf{x}) = p_{\mathbf{z}}(T^{-1}(\mathbf{x})) \, |\det J_{T^{-1}}(\mathbf{x})|. \qquad (13.2)$$

Now, we consider a chain of two blocks of invertible transformations, say $\mathbf{x} = T_2(T_1(\mathbf{z})) = T_2 \circ T_1(\mathbf{z})$. Clearly, $(T_2 \circ T_1)^{-1} = T_1^{-1} \circ T_2^{-1}$ and the Jacobian satisfies

$$\det J_{T_2 \circ T_1}(\mathbf{z}) = \det J_{T_2}(T_1(\mathbf{z})) \det J_{T_1}(\mathbf{z}). \qquad (13.3)$$

More generally, consider a chain of K blocks where $\mathbf{x} = T_K \circ \cdots \circ T_1(\mathbf{z})$. Let $\mathbf{z}_0 = \mathbf{z}$ and $\mathbf{z}_k = T_k(\mathbf{z}_{k-1})$ for $k = 1, \ldots, K$ (so that $\mathbf{x} = \mathbf{z}_K$). We have

$$\det J_{T_K \circ \cdots \circ T_1}(\mathbf{z}) = \prod_{k=1}^{K} \det J_{T_k}(\mathbf{z}_{k-1}). \qquad (13.4)$$

As a density model, normalizing flows can be used for both sampling and evaluating the probability density. When fitting normalizing flows to data, we want to minimize the discrepancy between the flow model (which expresses a distribution) and the target distribution. We already have seen many possible "discrepancy" metric that can be used in previous chapters. For instance, we can choose to minimize the KL divergence when training a flow.

Let $p_{\mathbf{x}}(\mathbf{x}; \boldsymbol{\theta})$ denote a normalizing flow model parameterized by $\boldsymbol{\theta}$, which can be decomposed into ϕ for the transformation T and ψ for the base distribution $p_{\mathbf{z}}$. Let $q(\mathbf{x})$ denote the groundtruth density. We can either choose to minimize $\mathrm{KL}(q \| p_{\mathbf{x}}(\cdot \, ; \boldsymbol{\theta}))$ or $\mathrm{KL}(p_{\mathbf{x}}(\cdot \, ; \boldsymbol{\theta}) \| q)$ with respect to $\boldsymbol{\theta}$. Since KL divergence is not symmetric in its arguments, these two settings result in different formulations. In particular,

$$\mathrm{KL}\left(q \| p_{\mathbf{x}}(\cdot \, ; \boldsymbol{\theta})\right) \qquad (13.5)$$

$$= \mathbb{E}_q[\log q - \log p_{\mathbf{x}}(\mathbf{x}; \boldsymbol{\theta})] \qquad (13.6)$$

$$= -\mathbb{E}_q[\log p_{\mathbf{x}}(\mathbf{x}; \boldsymbol{\theta})] - \mathrm{entropy}(q) \qquad (13.7)$$

$$= -\mathbb{E}_q\left[\log p_{\mathbf{z}}(T^{-1}(\mathbf{x}; \phi); \psi) + \log |\det J_{T^{-1}}(\mathbf{x}; \phi)|\right] + \mathrm{entropy}(q), \qquad (13.8)$$

where the last equality in (13.8) follows from (13.2). Also, $\mathrm{entropy}(q)$ is a constant independent of the parameters. Once we collect training examples $\{\mathbf{x}_n\}_{n=1}^N$, we can implement the Monte Carlo version of (13.8):

$$L(\boldsymbol{\phi}, \boldsymbol{\psi}) = -\frac{1}{N} \sum_{n=1}^N \log p_{\mathbf{z}}(T^{-1}(\mathbf{x}_n; \boldsymbol{\phi}); \boldsymbol{\psi}) + \log |\det J_{T^{-1}}(\mathbf{x}_n; \boldsymbol{\phi})|.$$

$$(13.9)$$

Note that the first term in (13.7) is simply the negative log likelihood of $p_{\mathbf{x}}(\mathbf{x}; \boldsymbol{\theta})$. Therefore, the loss function in (13.9) is also simply an implementation of maximum likelihood for the normalizing flow. When performing gradient descent, it is worth noting that the gradients of the loss function in (13.9) with respect to $\boldsymbol{\psi}$ only has to do with the first term.

In order to perform the above process, we need to have access to data sampled from the true distribution. In cases where it is difficult to sample from q, it is also useful consider the alternative

$$\mathrm{KL}\,(p_{\mathbf{x}}(\cdot\;;\boldsymbol{\theta})\|q) \tag{13.10}$$

$$= \mathbb{E}_{p_{\mathbf{x}}(\mathbf{x};\theta)}[\log p_{\mathbf{x}}(\mathbf{x}; \boldsymbol{\theta}) - \log q(\mathbf{x})] \tag{13.11}$$

$$= \mathbb{E}_{p_{\mathbf{z}}(\mathbf{z};\psi)}[\log p_{\mathbf{z}}(\mathbf{z}; \boldsymbol{\psi}) - \log |\det J_T(\mathbf{z}; \boldsymbol{\phi})| - \log q(T(\mathbf{z}; \boldsymbol{\phi}))]. \tag{13.12}$$

Even if we are not able to sample from q, it is possible that we can evaluate the density up to an unknown constant, say $q(\mathbf{x}) = \tilde{q}(\mathbf{x})/Z$, where Z is unknown. The multiplicative constant will become additive after taking the log. That is,

$$\mathrm{KL}\,(p_{\mathbf{x}}(\cdot\;;\boldsymbol{\theta})\|q) \tag{13.13}$$

$$= \mathbb{E}_{p_{\mathbf{z}}(\mathbf{z};\psi)}[\log p_{\mathbf{z}}(\mathbf{z}; \boldsymbol{\psi}) - \log |\det J_T(\mathbf{z}; \boldsymbol{\phi})| - \log \tilde{q}(T(\mathbf{z}; \boldsymbol{\phi}))]$$

$$+ \mathrm{const}. \tag{13.14}$$

When performing Monte Carlo, we only need to sample $\{\mathbf{z}_n\}_{n=1}^N$ following $p_{\mathbf{z}}(\mathbf{z}; \boldsymbol{\psi})$ and replace $\mathbb{E}_{p_{\mathbf{z}}(\mathbf{z};\psi)}$ with $\frac{1}{N}\sum_{n=1}^N$.

Fix a k which should not matter in the following discussion. In practice, we use neural networks to represent either T_k or T_k^{-1}. In the following, we use f_ϕ to denote such neural networks whose parameters are summarized in ϕ. In the following, we look into the *autoregressive flows*, which cover the most well-known classes of flows.

Let $\mathbf{z} = (z_1, \ldots, z_D)^\top$ denote an input feature. The entries $z_1, \ldots, z_D$ can be considered as a sequence. An autoregressive flow produces an output $\mathbf{z}' = (z_1', \ldots, z_D')^\top$ so that z_i' only depends on z_j for $j \leq i$, for each $i = 1, \ldots, D$. More specifically, the entries $z_1, \ldots, z_{i-1}$ are processed through a function c_i to produce a hidden feature $\mathbf{h}_i = c_i(\mathbf{z}_{1:i-1})$, where $\mathbf{z}_{1:i-1}$ denotes the $(i-1)$-dimensional vector $(z_1, \ldots, z_{i-1})^\top$. Then, z_i, together with $\mathbf{h}_i$, are processed by an invertible function τ, independent of the index i, to produce the output z_i'. That is, $z_i' = \tau(z_i; \mathbf{h}_i)$. In the work of Huang *et al.* (2018b), the function c_i is called a *conditioner* and τ is called a transformer. To avoid confusion with self-attention, we avoid using the name "transformer" for τ but call it a *synthesizer* instead.

Clearly, the inverse of the above autoregressive flow is given entriwise by $z_i = \tau^{-1}(z_i'; \mathbf{h}_i)$, where $\mathbf{h}_i = c_i(\mathbf{z}_{1:i-1})$. Note that given the fact that z_i' does not depend on z_j if $j > i$, the Jacobian of the flow is a lower-triangular matrix that can be written as

$$\frac{\partial \mathbf{z}'}{\partial \mathbf{z}} = \begin{bmatrix} \frac{\partial \tau}{\partial z_1}(z_1; \mathbf{h}_1) & & \mathbf{0} \\ & \ddots & \\ & \ddots & \frac{\partial \tau}{\partial z_D}(z_D; \mathbf{h}_D) \end{bmatrix}. \tag{13.15}$$

The determinant of the lower-triangular matrix is the produce of the diagonal entries. Let f_ϕ denote the function that represents the flow.

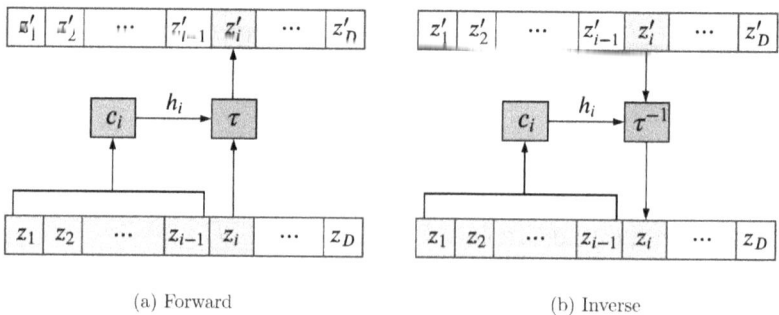

(a) Forward   (b) Inverse

Fig. 13.3. Autoregressive flow (Papamakarios *et al.*, 2021). The figure is taken from the paper under license: CC-BY 4.0. https://creativecommons.org/licenses/by/4.0/.

It is thus very convenient to derive that

$$\log \left| \det J_{f_\phi}(\mathbf{z}) \right| = \sum_{i=1}^{D} \log \left| \frac{\partial \tau}{\partial z_i}(z_i; \mathbf{h}_i) \right|. \tag{13.16}$$

The simplest examples of autoregressive flows are the ones that use affine functions as synthesizers. That is, $\tau(z_i; \mathbf{h}_i) = \alpha_i z_i + \beta_i$, where α_i and β_i compose $\mathbf{h}_i$, which is extracted from $\mathbf{z}_{1:i-1}$ by the conditioner c_i. To make τ invertible, we only need to ensure that $\alpha_i \neq 0$, which can be achieved e.g., if we set $\alpha_i = \exp(\tilde{\alpha}_i)$. In this case, $\frac{\partial \tau}{\partial z_i}(z_i; \mathbf{h}_i) = \frac{\partial}{\partial z_i}(\alpha_i z_i + \beta_i) = \alpha_i$ and thus

$$\log \left| \det J_{f_\phi}(\mathbf{z}) \right| = \sum_{i=1}^{D} \log |\alpha_i| = \sum_{i=1}^{D} \tilde{\alpha}_i. \tag{13.17}$$

One specific model where affine synthesizers are used is the real-valued non-volume-preserving (real NVP) flow. In this model, the so-called "coupling layers" are used. Specifically, fix an index d, e.g., $d = D/2$. The hidden features are constructed so that $\alpha_1, \ldots, \alpha_d = 1$ and $\beta_1, \ldots, \beta_d = 0$. In other words, $\mathbf{z}'_{1:d} = \mathbf{z}_{1:d}$. On the other hand, $\alpha_{d+1}, \ldots, \alpha_D$ and $\beta_{d+1}, \ldots, \beta_D$ only depend on $\mathbf{z}_{1:d}$ and do not depend on the indices $d+1, \ldots, D$. More concisely, we write

$$\mathbf{z}'_{1:d} = \mathbf{z}_{1:d}, \tag{13.18}$$

$$\mathbf{z}'_{d+1:D} = \mathbf{z}_{d+1:D} \odot \exp\left(\alpha(\mathbf{z}_{1:d})\right) + \beta(\mathbf{z}_{1:d}), \tag{13.19}$$

where α and β are conditioners for producing α_i's and β_i's, implemented by neural networks.

Problem 13.1.

(1) Write the inverse of the real NVP flow.
(2) Express the Jacobian matrix $\partial \mathbf{z}'/\partial \mathbf{z}$ for the real NVP flow.

Besides autoregressive flows, other types of normalizing flows include, e.g., linear flows and residual flows. In particular, residual flows are constructed similarly to residual networks (ResNets). However, unlike autoregressive flows, other types of flows usually do not have a very convenient form of Jacobian.

It is also possible to implement batch normalization between two flows. That is, $T_k \circ \text{BN} \circ T_{k-1}$. Nevertheless, the transformation of batch normalization is affine and thus it is easy to express the Jacobian. Specifically, let

$$\mathbf{z}' = \mathrm{BN}(\mathbf{z}) = \alpha \odot \frac{\mathbf{z} - \boldsymbol{\mu}}{\sqrt{\boldsymbol{\sigma}^2 + \boldsymbol{\epsilon}}} + \boldsymbol{\beta}. \tag{13.20}$$

We have

$$\log \left| \det \frac{\partial \mathbf{z}'}{\partial \mathbf{z}} \right| = \sum_{i=1}^{D} \log \frac{|\alpha_i|}{\sqrt{\sigma_i^2 + \epsilon_i}}. \tag{13.21}$$

13.2   Diffusion Models

One main benefit of normalizing flows is tractable representations of densities. However, when used as generative models, normalizing flows usually do not generate samples of the same quality as GANs. Diffusion models are another family of generative models that have recently aroused great interest since their capacity of generating samples of great quality.

In essence, a diffusion model can be regarded as a generalized VAE. In a VAE, the observable variable $\mathbf{x}$ is modeled by virtue of a latent variable $\mathbf{z}$. In a diffusion model, there is a chain of these latent variables (so that the structure of the model is similar to a normalizing flow). Specifically, if we use $\mathbf{x}_0$ to denote the observable variable, a diffusion model builds a chain $\mathbf{x}_1, \ldots, \mathbf{x}_T$ of latent variables, that satisfy the following conditions:

(1) The latent dimension is equal to the data dimension: $\dim(\mathbf{x}_t) = \dim(\mathbf{x}_0)$ for $t = 1, \ldots, T$.
(2) $\{\mathbf{x}_t\}_{t=0}^{T}$ is Markov and $q(\mathbf{x}_t|\mathbf{x}_{t-1})$ is a linear Gaussian: $q(\mathbf{x}_t|\mathbf{x}_{t-1}) = \mathcal{N}(\mathbf{x}_t|\sqrt{\alpha_t}\mathbf{x}_{t-1}, (1-\alpha_t)\mathbf{I})$, where α_t is a pre-defined scalar for each $t = 1, \ldots, T$.
(3) The final latent variable $\mathbf{x}_T$ follows a standard normal prior distribution: $p(\mathbf{x}_T) = \mathcal{N}(\mathbf{x}_T|\mathbf{0}, \mathbf{I})$.

The above conditions impose the rules for the *forward diffusion process*. In particular, at each time step t, we produce $\mathbf{x}_t$ by injecting a Gaussian noise to $\mathbf{x}_{t-1}$. We produce $\mathbf{x}_T$ after T time steps, which has *a priori* a standard normal distribution. The forward process is analogous to the encoder of a VAE. Similar to VAEs, if we want

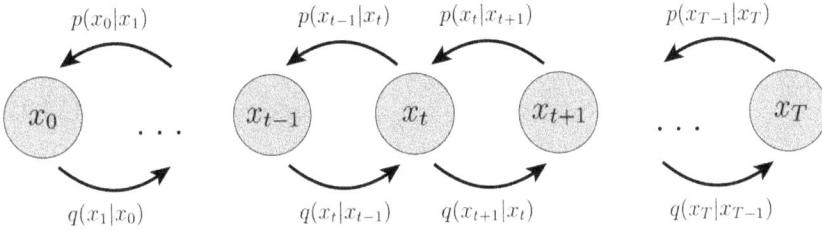

Fig. 13.4. Diffusion model.

samples that follow the same distribution of the input $\mathbf{x}_0$, we follow a *backward diffusion process*, which reverses the above noise injection process. That is why diffusion models are also called denoising diffusion models. The backward diffusion process is analogous to the decoder of a VAE. It is not hard to imagine the backward process is also intractable. Nevertheless, similar to VAE, we can derive an ELBO for a diffusion model, whose maximization will lead to a loss function for training the model, shown in Figure 13.4. In what follows, we describe the details of the forward process, the backward process as well as the training loss for the diffusion models.

13.2.1 *Forward diffusion process*

Given an input $\mathbf{x}_0$ that follows the data distribution, the forward diffusion process is specified by the conditional densities

$$q(\mathbf{x}_t|\mathbf{x}_{t-1}) = \mathcal{N}(\mathbf{x}_t|\sqrt{\alpha_t}\mathbf{x}_{t-1}, (1 - \alpha_t)\mathbf{I}). \qquad (13.22)$$

It is assumed that the forward process is Markov so that

$$q(\mathbf{x}_t|\mathbf{x}_{t-1}, \ldots, \mathbf{x}_0) = q(\mathbf{x}_t|\mathbf{x}_{t-1}). \qquad (13.23)$$

Therefore,

$$q(\mathbf{x}_{1:T}|\mathbf{x}_0) = \prod_{t=1}^{T} q(\mathbf{x}_t|\mathbf{x}_{t-1}, \ldots, \mathbf{x}_0) = \prod_{t=1}^{T} q(\mathbf{x}_t|\mathbf{x}_{t-1}). \qquad (13.24)$$

In (13.22), $\{\alpha_t\}_{t=1}^{T}$ is called the *variance schedule* (we abused the terminology since $\{1 - \alpha_t\}_{t=1}^{T}$ is the variance). A simple treatment is to treat α_t as a pre-defined parameter, though learning it is possible.

Note that governed by the conditional density (13.22), given $\mathbf{x}_{t-1}$, it is possible to sample $\mathbf{x}_t$ following the reparameterization regime:

$$\mathbf{x}_t = \sqrt{\alpha_t}\mathbf{x}_{t-1} + \sqrt{1 - \alpha_t}\epsilon_{t-1}, \tag{13.25}$$

where ϵ_{t-1} is a standard Gaussian noise: $p(\epsilon_{t-1}) = \mathcal{N}(\epsilon_{t-1}|\mathbf{0}, \mathbf{I})$. Clearly, (13.25) is a recursive relation which can be used to relate $\mathbf{x}_t$ to $\mathbf{x}_0$. We proceed as

$$\mathbf{x}_t = \sqrt{\alpha_t}\mathbf{x}_{t-1} + \sqrt{1 - \alpha_t}\epsilon_{t-1} \tag{13.26}$$

$$= \sqrt{\alpha_t}\left(\sqrt{\alpha_{t-1}}\mathbf{x}_{t-2} + \sqrt{1 - \alpha_{t-1}}\epsilon_{t-2}\right) + \sqrt{1 - \alpha_t}\epsilon_{t-1} \tag{13.27}$$

$$= \sqrt{\alpha_t\alpha_{t-1}}\mathbf{x}_{t-2} + \sqrt{\alpha_t - \alpha_t\alpha_{t-1}}\epsilon_{t-2} + \sqrt{1 - \alpha_t}\epsilon_{t-1}. \tag{13.28}$$

Before continuing, it is worth noting that the two uncorrelated noise terms in (13.28) can be realized and thus replaced by a single noise. The reason is that the sum of two uncorrelated Gaussian vectors is again a Gaussian vector. Since both ϵ_{t-2} and ϵ_{t-1} follow the standard normal distribution, the sum $\sqrt{\alpha_t - \alpha_t\alpha_{t-1}}\epsilon_{t-2} + \sqrt{1 - \alpha_t}\epsilon_{t-1}$ is also a Gaussian whose mean is the zero vector and whose variance of each entry is $(\sqrt{\alpha_t - \alpha_t\alpha_{t-1}})^2 + (\sqrt{1 - \alpha_t})^2 = 1 - \alpha_t\alpha_{t-1}$. By reparameterization, we have

$$\mathbf{x}_t = \sqrt{\alpha_t\alpha_{t-1}}\mathbf{x}_{t-2} + \sqrt{1 - \alpha_t\alpha_{t-1}}\epsilon_{t-2}, \tag{13.29}$$

where $p(\epsilon_{t-2}) = \mathcal{N}(\epsilon_{t-2}|\mathbf{0}, \mathbf{I})$. Note that we have abused the notation and call random noise ϵ_{t-2} again for simplicity. Continuing with this recursive relation, we can derive

$$\mathbf{x}_t = \sqrt{\bar{\alpha}_t}\mathbf{x}_0 + \sqrt{1 - \bar{\alpha}_t}\epsilon_0, \tag{13.30}$$

where $\bar{\alpha}_t = \prod_{i=1}^{t}\alpha_i$, and $p(\epsilon_0) = \mathcal{N}(\epsilon_0|\mathbf{0}, \mathbf{I})$. Clearly, (13.30) also reads

$$q(\mathbf{x}_t|\mathbf{x}_0) = \mathcal{N}(\mathbf{x}_t|\sqrt{\bar{\alpha}_t}\mathbf{x}_0, (1 - \bar{\alpha}_t)\mathbf{I}). \tag{13.31}$$

13.2.2 *Backward diffusion process*

In order to derive the backward diffusion process, we will seek a new process $p(\mathbf{x}_{t-1}|\mathbf{x}_t)$ as an approximation to the $q(\mathbf{x}_{t-1}|\mathbf{x}_t)$, which would govern the backward diffusion. By the last assumption of a diffusion model, the final latent variable $\mathbf{x}_T$ has a standard normal distribution. Even with this assumption, it is difficult to analytically derive the expression for $q(\mathbf{x}_{t-1}|\mathbf{x}_t)$. In order to so, we look at a the reverse process conditioned on the input $\mathbf{x}_0$. It is tractable now to calculate

$$q(\mathbf{x}_{t-1}|\mathbf{x}_t, \mathbf{x}_0) = \frac{q(\mathbf{x}_t|\mathbf{x}_{t-1}, \mathbf{x}_0)q(\mathbf{x}_{t-1}|\mathbf{x}_0)}{q(\mathbf{x}_t|\mathbf{x}_0)}. \tag{13.32}$$

Considering that $q(\mathbf{x}_{t-1}|\mathbf{x}_0)$ and $q(\mathbf{x}_t|\mathbf{x}_0)$ are clearly; and also, $q(\mathbf{x}_t|\mathbf{x}_{t-1}, \mathbf{x}_0) = q(\mathbf{x}_t|\mathbf{x}_{t-1})$ by the Markov assumption, which is also a linear Gaussian, we can derive the expression for input conditioned reverse process as a Gaussian by writing out the densities in (13.32) and then completing the squares to reveal the mean vector and the covariance matrix of $q(\mathbf{x}_{t-1}|\mathbf{x}_t, \mathbf{x}_0)$.

Alternatively, using a well-known fact for linear Gaussian models (Bishop and Nasrabadi, 2006, Chapter 2) that if $p(\mathbf{x}) = \mathcal{N}(\mathbf{x}|\boldsymbol{\mu}, \boldsymbol{\Lambda}^{-1})$ and $p(\mathbf{y}|\mathbf{x}) = \mathcal{N}(\mathbf{y}|\mathbf{Ax} + \mathbf{b}, \mathbf{L}^{-1})$, we derive

$$p(\mathbf{x}|\mathbf{y}) = \mathcal{N}(\mathbf{x}|\boldsymbol{\Sigma}(\mathbf{A}^\top \mathbf{L}(\mathbf{y} - \mathbf{b}) + \boldsymbol{\Lambda}\boldsymbol{\mu}), \boldsymbol{\Sigma}), \tag{13.33}$$

where $\boldsymbol{\Sigma} = (\boldsymbol{\Lambda} + \mathbf{A}^\top \mathbf{LA})^{-1}$. By substituting $\boldsymbol{\mu} = \sqrt{\bar{\alpha}_t}\mathbf{x}_0$, $\boldsymbol{\Lambda} = (1 - \bar{\alpha}_{t-1})^{-1}\mathbf{I}$, $\mathbf{A} = \sqrt{\alpha_t}\mathbf{I}$, $\mathbf{b} = \mathbf{0}$ and $\mathbf{L} = (1 - \alpha_t)^{-1}\mathbf{I}$, we can calculate

$$\boldsymbol{\Sigma} = \frac{(1 - \alpha_t)(1 - \bar{\alpha}_{t-1})}{1 - \bar{\alpha}_t}\mathbf{I} \tag{13.34}$$

and

$$\boldsymbol{\Sigma}(\mathbf{A}^\top \mathbf{L}(\mathbf{y} - \mathbf{b}) + \boldsymbol{\Lambda}\boldsymbol{\mu}) = \frac{\sqrt{\alpha_t}(1 - \bar{\alpha}_{t-1})}{1 - \bar{\alpha}_t}\mathbf{x}_t + \frac{\sqrt{\bar{\alpha}_{t-1}}(1 - \alpha_t)}{1 - \bar{\alpha}_t}\mathbf{x}_0. \tag{13.35}$$

That is, the density $q(\mathbf{x}_{t-1}|\mathbf{x}_t, \mathbf{x}_0)$ is given by

$$\mathcal{N}\left(\mathbf{x}_{t-1} \middle| \frac{\sqrt{\alpha_t}(1 - \bar{\alpha}_{t-1})}{1 - \bar{\alpha}_t}\mathbf{x}_t + \frac{\sqrt{\bar{\alpha}_{t-1}}(1 - \alpha_t)}{1 - \bar{\alpha}_t}\mathbf{x}_0, \frac{(1 - \alpha_t)(1 - \bar{\alpha}_{t-1})}{1 - \bar{\alpha}_t}\mathbf{I}\right). \tag{13.36}$$

Let's denote

$$\boldsymbol{\mu}_q(\mathbf{x}_t, \mathbf{x}_0) := \frac{\sqrt{\alpha_t}(1 - \bar{\alpha}_{t-1})}{1 - \bar{\alpha}_t}\mathbf{x}_t + \frac{\sqrt{\bar{\alpha}_{t-1}}(1 - \alpha_t)}{1 - \bar{\alpha}_t}\mathbf{x}_0 \qquad (13.37)$$

and

$$\boldsymbol{\Sigma}_q(t) := \frac{(1 - \alpha_t)(1 - \bar{\alpha}_{t-1})}{1 - \bar{\alpha}_t}\mathbf{I}. \qquad (13.38)$$

This gives us an expression $q(\mathbf{x}_{t-1}|\mathbf{x}_t, \mathbf{x}_0) = \mathcal{N}(\mathbf{x}_{t-1}|\boldsymbol{\mu}_q(\mathbf{x}_t, \mathbf{x}_0), \boldsymbol{\Sigma}_q(t))$ as a Gaussian whose mean depends on x_t and x_0 and a variance that depends on t. Although this expression governs the backward process only when conditioned on $\mathbf{x}_0$, it sheds light on the unconditioned process in the following two ways:

First, we can choose to model the backward $q(\mathbf{x}_{t-1}|\mathbf{x}_t)$ as a Gaussian whose covariance is the same as $\boldsymbol{\Sigma}_q(t)$ (there are various choices, but we keep with this choice for simplicity). We write the inverse diffusion as

$$p_{\boldsymbol{\theta}}(\mathbf{x}_{t-1}|\mathbf{x}_t) = \mathcal{N}\left(\mathbf{x}_{t-1}|\boldsymbol{\mu}_{\boldsymbol{\theta}}(\mathbf{x}_t; t), \boldsymbol{\Sigma}_q(t)\right), \qquad (13.39)$$

where $\boldsymbol{\mu}_{\boldsymbol{\theta}}(\cdot; t)$ is implemented by a neural network.

Next, we approximate the mean by learning $\mu_{\boldsymbol{\theta}}(\cdot; t)$ using ELBO.

13.2.3　*Variational inference*

Let's derive the ELBO for the diffusion model and we find that $q(\mathbf{x}_{t-1}|\mathbf{x}_t, \mathbf{x}_0)$ plays an important role. In particular, to get a lower bound for $\log p(\mathbf{x}_0)$, we consider $\mathbf{x}_{1:T}$ as a joint latent variable. The corresponding ELBO reads

$$\log p(\mathbf{x}_0) \geq \mathbb{E}_{q(\mathbf{x}_{1:T}|\mathbf{x}_0)}\left[\log \frac{p(\mathbf{x}_{0:T})}{q(\mathbf{x}_{1:T}|\mathbf{x}_0)}\right]. \qquad (13.40)$$

To make more sense of the right-hand side, we proceed as follows:

$$\log p(\mathbf{x}_0) \geq \mathbb{E}_{q(\mathbf{x}_{1:T}|\mathbf{x}_0)}\left[\log \frac{p(\mathbf{x}_{0:T})}{q(\mathbf{x}_{1:T}|\mathbf{x}_0)}\right]$$

$$= \mathbb{E}_{q(\mathbf{x}_{1:T}|\mathbf{x}_0)}\left[\log \frac{p(\mathbf{x}_T)p_{\boldsymbol{\theta}}(\mathbf{x}_0|\mathbf{x}_1)\prod_{t=2}^{T} p_{\boldsymbol{\theta}}(\mathbf{x}_{t-1}|\mathbf{x}_t)}{q(\mathbf{x}_1|\mathbf{x}_0)\prod_{t=2}^{T} q(\mathbf{x}_t|\mathbf{x}_{t-1})}\right]$$

$$\qquad (13.41)$$

$$= \mathbb{E}_{q(\mathbf{x}_{1:T}|\mathbf{x}_0)} \left[\log \frac{p(\mathbf{x}_T) p_{\boldsymbol{\theta}}(\mathbf{x}_0|\mathbf{x}_1) \prod_{t=2}^{T} p_{\boldsymbol{\theta}}(\mathbf{x}_{t-1}|\mathbf{x}_t)}{q(\mathbf{x}_1|\mathbf{x}_0) \prod_{t=2}^{T} q(\mathbf{x}_t|\mathbf{x}_{t-1}, \mathbf{x}_0)} \right] \tag{13.42}$$

$$= \mathbb{E}_{q(\mathbf{x}_{1:T}|\mathbf{x}_0)} \left[\log \frac{p_{\boldsymbol{\theta}}(\mathbf{x}_T) p_{\boldsymbol{\theta}}(\mathbf{x}_0|\mathbf{x}_1)}{q(\mathbf{x}_1|\mathbf{x}_0)} + \log \prod_{t=2}^{T} \frac{p_{\boldsymbol{\theta}}(\mathbf{x}_{t-1}|\mathbf{x}_t)}{q(\mathbf{x}_t|\mathbf{x}_{t-1}, \mathbf{x}_0)} \right], \tag{13.43}$$

where in (13.42) we deliberately condition $q(\mathbf{x}_t|\mathbf{x}_{t-1})$ on $\mathbf{x}_0$, which has no effect because of the Markov assumption. This facilitates the use of (13.32) and we continue as follows:

$$\log p(\mathbf{x}_0)$$

$$\geq \mathbb{E}_{q(\mathbf{x}_{1:T}|\mathbf{x}_0)} \left[\log \frac{p(\mathbf{x}_T) p_{\boldsymbol{\theta}}(\mathbf{x}_0|\mathbf{x}_1)}{q(\mathbf{x}_1|\mathbf{x}_0)} + \log \prod_{t=2}^{T} \frac{p_{\boldsymbol{\theta}}(\mathbf{x}_{t-1}|\mathbf{x}_t)}{\frac{q(\mathbf{x}_{t-1}|\mathbf{x}_t, \mathbf{x}_0) q(\mathbf{x}_t|\mathbf{x}_0)}{q(\mathbf{x}_{t-1}|\mathbf{x}_0)}} \right] \tag{13.44}$$

$$= \mathbb{E}_{q(\mathbf{x}_{1:T}|\mathbf{x}_0)} \left[\log \frac{p(\mathbf{x}_T) p_{\boldsymbol{\theta}}(\mathbf{x}_0|\mathbf{x}_1)}{q(\mathbf{x}_1|\mathbf{x}_0)} + \log \frac{q(\mathbf{x}_1|\mathbf{x}_0)}{q(\mathbf{x}_T|\mathbf{x}_0)} \right.$$

$$\left. + \log \prod_{t=2}^{T} \frac{p_{\boldsymbol{\theta}}(\mathbf{x}_{t-1}|\mathbf{x}_t)}{q(\mathbf{x}_{t-1}|\mathbf{x}_t, \mathbf{x}_0)} \right] \tag{13.45}$$

$$= \mathbb{E}_{q(\mathbf{x}_{1:T}|\mathbf{x}_0)} \left[\log \frac{p(\mathbf{x}_T) p_{\boldsymbol{\theta}}(\mathbf{x}_0|\mathbf{x}_1)}{q(\mathbf{x}_T|\mathbf{x}_0)} + \log \prod_{t=2}^{T} \frac{p_{\boldsymbol{\theta}}(\mathbf{x}_{t-1}|\mathbf{x}_t)}{q(\mathbf{x}_{t-1}|\mathbf{x}_t, \mathbf{x}_0)} \right] \tag{13.46}$$

$$= \mathbb{E}_{q(\mathbf{x}_{1:T}|\mathbf{x}_0)} \left[\log p_{\boldsymbol{\theta}}(\mathbf{x}_0|\mathbf{x}_1) \right] + \mathbb{E}_{q(\mathbf{x}_{1:T}|\mathbf{x}_0)} \left[\log \frac{p(\mathbf{x}_T)}{q(\mathbf{x}_T|\mathbf{x}_0)} \right]$$

$$+ \sum_{t=2}^{T} \mathbb{E}_{q(\mathbf{x}_{1:T}|\mathbf{x}_0)} \left[\log \frac{p_{\boldsymbol{\theta}}(\mathbf{x}_{t-1}|\mathbf{x}_t)}{q(\mathbf{x}_{t-1}|\mathbf{x}_t, \mathbf{x}_0)} \right] \tag{13.47}$$

$$= \mathbb{E}_{q(\mathbf{x}_1|\mathbf{x}_0)} \left[\log p_{\boldsymbol{\theta}}(\mathbf{x}_0|\mathbf{x}_1) \right] + \mathbb{E}_{q(\mathbf{x}_T|\mathbf{x}_0)} \left[\log \frac{p(\mathbf{x}_T)}{q(\mathbf{x}_T|\mathbf{x}_0)} \right]$$

$$+ \sum_{t=2}^{T} \mathbb{E}_{q(\mathbf{x}_{t-1}|\mathbf{x}_t, \mathbf{x}_0)} \left[\log \frac{p_{\boldsymbol{\theta}}(\mathbf{x}_{t-1}|\mathbf{x}_t)}{q(\mathbf{x}_{t-1}|\mathbf{x}_t, \mathbf{x}_0)} \right]. \tag{13.48}$$

13.2.4 *Loss function*

Clearly, the second term of (13.48) is the KL divergence from $q(\mathbf{x}_T|\mathbf{x}_0)$ to $p(\mathbf{x}_T)$ and there are no parameters to learn in this term. The first term is similar to the reconstruction term in a VAE. Therefore, the last summation term is the most interesting. To proceed, we can write $\mathbb{E}_{q(\mathbf{x}_{t-1}|\mathbf{x}_t,\mathbf{x}_0)}\left[\log\frac{p_\theta(\mathbf{x}_{t-1}|\mathbf{x}_t)}{q(\mathbf{x}_{t-1}|\mathbf{x}_t,\mathbf{x}_0)}\right]$ as

$$- \mathrm{KL}\left(q(\mathbf{x}_{t-1}|\mathbf{x}_t,\mathbf{x}_0)\|p_\theta(\mathbf{x}_{t-1}|\mathbf{x}_t)\right). \tag{13.49}$$

Note that both arguments of the KL divergence are Gaussian distributions, whose covariance matrices are the same by our assumption. We first state a lemma that calculates the KL divergence of generic Gaussians, which is of its own interest.

Lemma 13.1. *Let $p_1(\mathbf{x}) = \mathcal{N}(\mathbf{x}|\boldsymbol{\mu}_1,\boldsymbol{\Sigma}_1)$ and $p_2(\mathbf{x}) = \mathcal{N}(\mathbf{x}|\boldsymbol{\mu}_2,\boldsymbol{\Sigma}_2)$ be two Gaussian distributions in $\mathbb{R}^D$. Then,*

$$\mathrm{KL}\left(p_1\|p_2\right)$$
$$= \frac{1}{2}\left(\log\frac{\det\boldsymbol{\Sigma}_2}{\det\boldsymbol{\Sigma}_1} - D + \mathrm{tr}(\boldsymbol{\Sigma}_2^{-1}\boldsymbol{\Sigma}_1) + (\boldsymbol{\mu}_2-\boldsymbol{\mu}_1)^\top\boldsymbol{\Sigma}_2^{-1}(\boldsymbol{\mu}_2-\boldsymbol{\mu}_1)\right). \tag{13.50}$$

Proof.   Note that for $j = 1,2$,

$$p_j(\mathbf{x}) = \frac{1}{(2\pi)^{D/2}(\det\boldsymbol{\Sigma}_j)^{1/2}}\exp\left(-\frac{1}{2}(\mathbf{x}-\boldsymbol{\mu}_j)^\top\boldsymbol{\Sigma}_j^{-1}(\mathbf{x}-\boldsymbol{\mu}_j)\right). \tag{13.51}$$

Accordingly,

$$\mathrm{KL}\left(p_1\|p_2\right) := \mathbb{E}_{p_1}\left[\log p_1(\mathbf{x}) - \log p_2(\mathbf{x})\right] \tag{13.52}$$

$$= \mathbb{E}_{p_1}\left[\frac{1}{2}\log\frac{\det\boldsymbol{\Sigma}_2}{\det\boldsymbol{\Sigma}_1} - \frac{1}{2}(\mathbf{x}-\boldsymbol{\mu}_1)^\top\boldsymbol{\Sigma}_1^{-1}(\mathbf{x}-\boldsymbol{\mu}_1)\right.$$
$$\left.+ \frac{1}{2}(\mathbf{x}-\boldsymbol{\mu}_2)^\top\boldsymbol{\Sigma}_2^{-1}(\mathbf{x}-\boldsymbol{\mu}_2)\right] \tag{13.53}$$

$$= \frac{1}{2} \log \frac{\det \boldsymbol{\Sigma}_2}{\det \boldsymbol{\Sigma}_1} - \frac{1}{2} \mathbb{E}_{\mathcal{N}(\mathbf{x}|0,\boldsymbol{\Sigma}_1)} \left[\mathbf{x}^\top \boldsymbol{\Sigma}_1^{-1} \mathbf{x} \right]$$

$$+ \frac{1}{2} \mathbb{E}_{\mathcal{N}(\mathbf{x}|0,\boldsymbol{\Sigma}_1)} \left[(\mathbf{x} - \boldsymbol{\mu}_2 + \boldsymbol{\mu}_1)^\top \boldsymbol{\Sigma}_2^{-1} (\mathbf{x} - \boldsymbol{\mu}_2 + \boldsymbol{\mu}_1) \right],$$

$$(13.54)$$

where the equality of (13.54) follows a change of variable $\mathbf{x} \mapsto \mathbf{x} + \boldsymbol{\mu}_1$. We proceed by calculating the second and third terms in (13.54) as follows. For the second term,

$$\frac{1}{2} \mathbb{E}_{\mathcal{N}(\mathbf{x}|0,\boldsymbol{\Sigma}_1)} \left[\mathbf{x}^\top \boldsymbol{\Sigma}_1^{-1} \mathbf{x} \right]$$

$$= \frac{1}{2} \int \mathbf{x}^\top \boldsymbol{\Sigma}_1^{-1} \mathbf{x} \frac{1}{(2\pi)^{D/2} (\det \boldsymbol{\Sigma}_1)^{1/2}} \exp\left(-\frac{1}{2} \mathbf{x}^\top \boldsymbol{\Sigma}_1^{-1} \mathbf{x} \right) d\mathbf{x}$$

$$(13.55)$$

$$= \frac{1}{2} \int \|\mathbf{z}\|^2 \frac{1}{(2\pi)^{D/2}} \exp\left(-\frac{1}{2} \|\mathbf{z}\|^2 \right) d\mathbf{z} \tag{13.56}$$

$$= \frac{1}{2} D, \tag{13.57}$$

where the equality of (13.57) is immediate once we write $\|\mathbf{z}\|^2 = \sum_{i=1}^{D} z_i^2$ and regard (13.56) as its expectation with respect to the standard normal distribution. For the third term,

$$\frac{1}{2} \mathbb{E}_{\mathcal{N}(\mathbf{x}|0,\boldsymbol{\Sigma}_1)} \left[(\mathbf{x} - \boldsymbol{\mu}_2 + \boldsymbol{\mu}_1)^\top \boldsymbol{\Sigma}_2^{-1} (\mathbf{x} - \boldsymbol{\mu}_2 + \boldsymbol{\mu}_1) \right]$$

$$= \frac{1}{2} \int (\mathbf{x} - \boldsymbol{\mu}_2 + \boldsymbol{\mu}_1)^\top \boldsymbol{\Sigma}_2^{-1} (\mathbf{x} - \boldsymbol{\mu}_2 + \boldsymbol{\mu}_1) \exp\left(-\frac{1}{2} \mathbf{x}^\top \boldsymbol{\Sigma}_1^{-1} \mathbf{x} \right) d\mathbf{x}$$

$$(13.58)$$

$$= \frac{1}{2} \int (\boldsymbol{\mu}_2 - \boldsymbol{\mu}_1)^\top \boldsymbol{\Sigma}_2^{-1} (\boldsymbol{\mu}_2 - \boldsymbol{\mu}_1) \exp\left(-\frac{1}{2} \mathbf{x}^\top \boldsymbol{\Sigma}_1^{-1} \mathbf{x} \right) d\mathbf{x}$$

$$- \int \mathbf{x}^\top \boldsymbol{\Sigma}_2^{-1} (\boldsymbol{\mu}_2 - \boldsymbol{\mu}_1) \exp\left(-\frac{1}{2} \mathbf{x}^\top \boldsymbol{\Sigma}_1^{-1} \mathbf{x} \right) d\mathbf{x}$$

$$+ \frac{1}{2} \int \mathbf{x}^\top \boldsymbol{\Sigma}_2^{-1} \mathbf{x} \exp\left(-\frac{1}{2} \mathbf{x}^\top \boldsymbol{\Sigma}_1^{-1} \mathbf{x} \right) d\mathbf{x} \tag{13.59}$$

$$= \frac{1}{2}(\boldsymbol{\mu}_2 - \boldsymbol{\mu}_1)^\top \boldsymbol{\Sigma}_2^{-1}(\boldsymbol{\mu}_2 - \boldsymbol{\mu}_1) - 0$$

$$+ \frac{1}{2} \int \mathbf{x}^\top \boldsymbol{\Sigma}_2^{-1} \mathbf{x} \exp\left(-\frac{1}{2}\mathbf{x}^\top \boldsymbol{\Sigma}_1^{-1}\mathbf{x}\right) d\mathbf{x} \tag{13.60}$$

$$= \frac{1}{2}(\boldsymbol{\mu}_2 - \boldsymbol{\mu}_1)^\top \boldsymbol{\Sigma}_2^{-1}(\boldsymbol{\mu}_2 - \boldsymbol{\mu}_1)$$

$$+ \frac{1}{2} \int \mathbf{z}^\top \boldsymbol{\Sigma}_1^{1/2} \boldsymbol{\Sigma}_2^{-1} \boldsymbol{\Sigma}_1^{1/2} \mathbf{z} \frac{1}{(2\pi)^{D/2}} \exp\left(-\frac{1}{2}\|\mathbf{z}\|^2\right) d\mathbf{z} \tag{13.61}$$

$$= \frac{1}{2}(\boldsymbol{\mu}_2 - \boldsymbol{\mu}_1)^\top \boldsymbol{\Sigma}_2^{-1}(\boldsymbol{\mu}_2 - \boldsymbol{\mu}_1) + \mathrm{tr}(\boldsymbol{\Sigma}_1^{1/2}\boldsymbol{\Sigma}_2^{-1}\boldsymbol{\Sigma}_1^{1/2}) \tag{13.62}$$

$$= \frac{1}{2}(\boldsymbol{\mu}_2 - \boldsymbol{\mu}_1)^\top \boldsymbol{\Sigma}_2^{-1}(\boldsymbol{\mu}_2 - \boldsymbol{\mu}_1) + \mathrm{tr}(\boldsymbol{\Sigma}_2^{-1}\boldsymbol{\Sigma}_1), \tag{13.63}$$

where (13.62) follows from a spectral decomposition of $\boldsymbol{\Sigma}_1^{1/2}\boldsymbol{\Sigma}_2^{-1}\boldsymbol{\Sigma}_1^{1/2}$ in (13.61) and taking expectation of the sum of eigenvalues multiplied with the squared variable with respect to a standard normal distribution, which is exactly the sum of eigenvalues, or the trace. We obtain (13.50) by substituting the second and third terms in (13.54) with (13.57) and (13.63), respectively. □

With the above lemma, we calculate

$$\mathrm{KL}\left(q(\mathbf{x}_t, \mathbf{x}_{t-1}|\mathbf{x}_0)\|p_{\boldsymbol{\theta}}(\mathbf{x}_{t-1}|\mathbf{x}_t)\right)$$

$$= \frac{1}{2}\left(0 - D + D + (\boldsymbol{\mu}_{\boldsymbol{\theta}} - \boldsymbol{\mu}_q)^\top \boldsymbol{\Sigma}_q^{-1}(\boldsymbol{\mu}_{\boldsymbol{\theta}} - \boldsymbol{\mu}_q)\right) \tag{13.64}$$

$$- \frac{1}{2}(\boldsymbol{\mu}_{\boldsymbol{\theta}} - \boldsymbol{\mu}_q)^\top \boldsymbol{\Sigma}_q^{-1}(\boldsymbol{\mu}_{\boldsymbol{\theta}} - \boldsymbol{\mu}_q), \tag{13.65}$$

where we recall that $\boldsymbol{\mu}_q$ and $\boldsymbol{\Sigma}_q$ are given by (13.37) and (13.38), respectively. Given the specific form in (13.37), it is convenient to impose a similar form for $\boldsymbol{\mu}_{\boldsymbol{\theta}}$. Specifically, we can define

$$\boldsymbol{\mu}_{\boldsymbol{\theta}}(\mathbf{x}_t; t) = \frac{\sqrt{\alpha_t}(1 - \bar{\alpha}_{t-1})}{1 - \bar{\alpha}_t}\mathbf{x}_t + \frac{\sqrt{\bar{\alpha}_{t-1}}(1 - \alpha_t)}{1 - \bar{\alpha}_t}\mathbf{x}_{\boldsymbol{\theta}}(\mathbf{x}_t; t). \tag{13.66}$$

Here, we only need a neural network representation of $\mathbf{x}_{\boldsymbol{\theta}}$. Then,

$$\boldsymbol{\mu}_{\boldsymbol{\theta}}(\mathbf{x}_t; t) - \boldsymbol{\mu}_q(\mathbf{x}_t, \mathbf{x}_0) = \frac{\sqrt{\bar{\alpha}_{t-1}}(1 - \alpha_t)}{1 - \bar{\alpha}_t}\left(\mathbf{x}_{\boldsymbol{\theta}}(\mathbf{x}_t; t) - \mathbf{x}_0\right) \tag{13.67}$$

and thus, following (13.65),

$$\mathrm{KL}\left(q(\mathbf{x}_t, \mathbf{x}_{t-1}|\mathbf{x}_0)\|p_{\boldsymbol{\theta}}(\mathbf{x}_{t-1}|\mathbf{x}_t)\right)$$

$$= \frac{1}{2}\frac{1-\bar{\alpha}_t}{(1-\alpha_t)(1-\bar{\alpha}_{t-1})}\frac{\bar{\alpha}_{t-1}(1-\alpha_t)^2}{(1-\bar{\alpha}_t)^2}\|\mathbf{x}_{\boldsymbol{\theta}}(\mathbf{x}_t; t) - \mathbf{x}_0\|^2 \quad (13.68)$$

$$= \frac{1}{2}\frac{\bar{\alpha}_{t-1}(1-\alpha_t)}{(1-\bar{\alpha}_{t-1})(1-\bar{\alpha}_t)}\|\mathbf{x}_{\boldsymbol{\theta}}(\mathbf{x}_t; t) - \mathbf{x}_0\|^2 \quad (13.69)$$

$$= \frac{1}{2}\frac{\bar{\alpha}_{t-1}-\bar{\alpha}_t}{(1-\bar{\alpha}_{t-1})(1-\bar{\alpha}_t)}\|\mathbf{x}_{\boldsymbol{\theta}}(\mathbf{x}_t; t) - \mathbf{x}_0\|^2. \quad (13.70)$$

Therefore, maximizing each summand in the third term of the ELBO expression (13.48) is equivalent to minimizing a reconstruction loss where $\mathbf{x}_{\boldsymbol{\theta}}(\mathbf{x}_t; t)$ seeks to predict the initial data point x_0 from the noisy version x_t using a neural network parameterized by $\boldsymbol{\theta}$. Instead of taking $\sum_{t=2}^{T}$ over all time steps t, we can sample time steps uniformly for efficient implementation. This is the first view of the diffusion problem that amounts to learning a neural network that predicts the original ground truth $\mathbf{x}_0$ from an arbitrary noisy version sampled at noise step t.

Additional intuition can be gained by rewriting the diffusion problem as a noise prediction problem, as follows. In the above, we have reduced the KL divergence to reconstruction loss for the input $\mathbf{x}_0$. Let's look at an alternative view. Recall from (13.30) that $\mathbf{x}_t = \sqrt{\bar{\alpha}_t}\mathbf{x}_0 + \sqrt{1-\bar{\alpha}_t}\boldsymbol{\epsilon}_0$. We can rearrange the terms to obtain

$$\mathbf{x}_0 = \frac{\mathbf{x}_t - \sqrt{1-\bar{\alpha}_t}\boldsymbol{\epsilon}_0}{\sqrt{\bar{\alpha}_t}}. \quad (13.71)$$

Together with (13.37) and some easy algebra, we have

$$\boldsymbol{\mu}_q(\mathbf{x}_t, \mathbf{x}_0) = \frac{1}{\sqrt{\alpha_t}}\mathbf{x}_t - \frac{1-\alpha_t}{\sqrt{1-\bar{\alpha}_t}\sqrt{\alpha_t}}\boldsymbol{\epsilon}_0. \quad (13.72)$$

With this formula, in view of (13.65), it is more feasible than (13.66) to set

$$\boldsymbol{\mu}_{\boldsymbol{\theta}}(\mathbf{x}_t; t) = \frac{1}{\sqrt{\alpha_t}}\mathbf{x}_t - \frac{1-\alpha_t}{\sqrt{1-\bar{\alpha}_t}\sqrt{\alpha_t}}\boldsymbol{\epsilon}_{\boldsymbol{\theta}}(\mathbf{x}_t; t), \quad (13.73)$$

where $\epsilon_\theta(\mathbf{x}_t; t)$ is a neural network that predicts the noise ϵ_0. Clearly,

$$\boldsymbol{\mu_\theta}(\mathbf{x}_t; t) - \boldsymbol{\mu_q}(\mathbf{x}_t, \mathbf{x}_0) = -\frac{1 - \alpha_t}{\sqrt{1 - \bar{\alpha}_t}\sqrt{\alpha_t}}(\boldsymbol{\epsilon_\theta}(\mathbf{x}_t; t) - \boldsymbol{\epsilon_0}). \quad (13.74)$$

The corresponding KL divergence is

$$\mathrm{KL}\left(q(\mathbf{x}_t, \mathbf{x}_{t-1}|\mathbf{x}_0)\|p_{\boldsymbol\theta}(\mathbf{x}_{t-1}|\mathbf{x}_t)\right)$$

$$= \frac{1}{2}\frac{1 - \bar{\alpha}_t}{(1 - \alpha_t)(1 - \bar{\alpha}_{t-1})}\frac{(1 - \alpha_t)^2}{(1 - \bar{\alpha}_t)\alpha_t}\|\boldsymbol{\epsilon_\theta}(\mathbf{x}_t; t) - \boldsymbol{\epsilon_0}\|^2 \quad (13.75)$$

$$= \frac{1}{2}\frac{1 - \alpha_t}{(1 - \bar{\alpha}_{t-1})\alpha_t}\|\boldsymbol{\epsilon_\theta}(\mathbf{x}_t; t) - \boldsymbol{\epsilon_0}\|^2. \quad (13.76)$$

In practice, it is simpler to implement an unweighted version of (13.76), which ignores the coefficient in terms of the variance schedule. Indeed, to facilitate easy training, we can employ the following simple loss function:

$$\ell_{\mathrm{simple}}(\boldsymbol{\theta}) = \mathbb{E}_{t,\mathbf{x}_0,\boldsymbol{\epsilon}_0}\|\boldsymbol{\epsilon_\theta}(\mathbf{x}_t; t) - \boldsymbol{\epsilon_0}\|^2, \quad (13.77)$$

where t is uniformly sampled from $\{1, \ldots, T\}$, $\mathbf{x}_0$ is the input sampled from data, and $\boldsymbol{\epsilon}_0$ is the noise sampled from $\mathcal{N}(\mathbf{0}, \mathbf{I})$, which is used to produce $\mathbf{x}_t$ according to (13.30).

This leads to the following realization of the forward step as a network learning to predict the noise independently at each of the noise steps, repeating the optimization process by randomly selecting the step and sampling of the noise.

Algorithm 1 Training

1: **repeat**
2:     $\mathbf{x}_0 \sim q(\mathbf{x}_0)$
3:     $t \sim \mathrm{Uniform}(\{1, \ldots, T\})$
4:     $\boldsymbol{\epsilon} \sim \mathcal{N}(\mathbf{0}, \mathbf{I})$
5:     Take gradient descent step on
    $\nabla_\theta\|\boldsymbol{\epsilon} - \boldsymbol{\epsilon_\theta}(\sqrt{\bar{\alpha}_t}x_0 + \sqrt{1 - \bar{\alpha}_t}\boldsymbol{\epsilon}, t)\|^2$
6: **until** converged

The backward process starts generates a data point by sampling it from the learned approximation of the distribution of $\mathbf{x}_0$. This sampling is done by going through multiple successive backward steps, where at each step an intermediate (latent) $\mathbf{x}_t$ is sampled according to the inverse diffusion process (13.39) using the reparametrization trick with estimated mean $\boldsymbol{\mu}_\theta(\mathbf{x}_t; t)$ (13.73) and variance schedule $\boldsymbol{\Sigma}_q(t)$ (13.38).

Algorithm 2 Sampling

1: $\mathbf{x}_T \sim \mathcal{N}(\mathbf{0}, \mathbf{I})$
2: **for** $t = T, \ldots, 1$ **do**
3: $\quad$ $\mathbf{z} \sim \mathcal{N}(\mathbf{0}, \mathbf{I})$ if $t > 1$, else $\mathbf{z} = \mathbf{0}$
4: $\quad$ $\mathbf{x}_{t-1} = \frac{1}{\sqrt{\alpha_t}} \left(x_t - \frac{1-\alpha_t}{\sqrt{1-\bar{\alpha}_t}} \boldsymbol{\epsilon}_\theta(\mathbf{x}_t, t) \right) + \boldsymbol{\Sigma}_q(t)\mathbf{z}$
5: **end for**
6: **return** x_0

The architecture of $\boldsymbol{\epsilon}_\theta$ can vary, and different choices have been explored in the literature. For example, a U-Net architecture has been utilized in the work of Dhariwal and Nichol (2021). As we have discussed in Chapter 8, the U-Net architecture is well known for its effective denoising capabilities, which is why it also demonstrates success in the denoising process within diffusion models.

13.2.5 *DDIM: Deterministic generative process*

A key drawback of denoising diffusion probabilistic models (DDPMs) is the need to simulate a lengthy Markov chain to generate a sample, which can be computationally expensive. To address this limitation, researchers developed denoising diffusion implicit models (DDIMs). DDIMs offer a more efficient approach by generalizing DDPMs through non-Markovian diffusion processes, enabling the use of deterministic generative processes that produce high-quality samples much faster. Both DDIM and DDPM are trained with the same objective function: minimizing $L_\gamma(\epsilon_\theta)$. The key difference lies in the underlying inference or forward model assumption, which leads to a new sampling procedure. DDPM employs a Markovian forward process with a Gaussian transition kernel that determines the distribution of x_t given x_{t-1}. $q(x_t|x_{t-1}) := N(\sqrt{\alpha_t \alpha_{t-1}} x_{t-1}, (1 - \frac{\alpha_t}{\alpha_{t-1}})I)$. The parameter α_t controls the amount of noise added at each time step. DDIM, on the other hand, utilizes a non-Markovian forward process where the sample at a given time step depends on both the noisy sample at the previous time step and the original data sample (x_0). This dependence structure is given by $q_\sigma(x_t|x_{t-1}, x_0) = \frac{q_\sigma(x_{t-1}|x_t, x_0) q_\sigma(x_t|x_0)}{q_\sigma(x_{t-1}|x_0)}$, where $q_\sigma(x_{t-1}|x_t, x_0) = N(\sqrt{\alpha_{t-1}} x_0 + \sqrt{1 - \alpha_{t-1} - \sigma_t^2} \cdot \frac{x_t - \sqrt{\alpha_t} x_0}{\sqrt{1-\alpha_t}}, \sigma_t^2 I)$ and $q_\sigma(x_t|x_0) = N(\sqrt{\alpha_t} x_0, (1 - \alpha_t) I)$. The parameter σ_t in the DDIM forward process controls the degree of stochasticity. When σ_t is set to $\sqrt{(1 - \alpha_{t-1})/(1 - \alpha_t)} \sqrt{1 - \alpha_t/\alpha_{t-1}}$, the forward process becomes equivalent to the Markovian process used in DDPM. However, DDIM allows for exploration of non-Markovian forward processes by setting σ_t to different values. This flexibility, combined with the deterministic nature of its sampling procedure, contributes to the improved sample efficiency and consistency observed in DDIM.

13.2.6 *DDIM sampling equation*

The DDIM sampling process (also known as Reverse Process) is given by

$$x_{t-1} = \sqrt{\alpha_{t-1}} \left(\frac{x_t - \sqrt{1 - \alpha_t} \epsilon_\theta^{(t)}(x_t)}{\sqrt{\alpha_t}} \right)$$

$$+ \sqrt{1 - \alpha_{t-1} - \sigma_t^2} \cdot \epsilon_\theta^{(t)}(x_t) + \sigma_t \epsilon_t, \tag{13.78}$$

where $\epsilon_t \sim N(0, I)$ is standard Gaussian noise, σ_t controls the stochasticity of the generative process, and $\epsilon_\theta^{(t)}(x_t)$ is the model that prodicts c_t given x_t.

When $\sigma_t = 0$ for all t, the sampling process becomes deterministic, resulting in the DDIM model.

The key advantages of DDIM over DDPM are sample efficiency and sample consistency. DDIM requires significantly fewer steps to produce high-quality samples compared to DDPM. It can achieve sample quality comparable to a 1000-step DDPM within 20–100 steps. Since the same initial latent variable x_T will generate samples with similar high-level features, this property is known as consistency. This allows DDIM to have semantically meaningful interpolation directly in the latent space by manipulating x_T, improving upon interpretability of the model.

13.2.7 *Score-matching approach*

The original motivation, and thus the name for diffusion models came from physics of diffusion processes, described by Langevin dynamics. We have seen above that the sampling process of DDPM (and DDIM for $\sigma_t > 0$) are using Langevin equation during inference. In the original physic formulation, the probability distribution of the reverse process is written using an energy function $f_\theta(\mathbf{x})$

$$p_\theta(\mathbf{x}) = \frac{1}{Z_\theta} e^{-f_\theta(\mathbf{x})}, \tag{13.79}$$

with Z_θ being a normalizing constant to ensure that $\int p_\theta(\mathbf{x})d\mathbf{x} = 1$. A common difficulty in learning such distributions is computing the normalizing constant $Z_\theta = \int e^{-f_\theta(\mathbf{x})}d\mathbf{x}$. This difficulty can be avoided by taking the derivative of the log of both sides

$$\nabla_\mathbf{x} \log p_\theta(\mathbf{x}) = \nabla_\mathbf{x} \log \left(\frac{1}{Z_\theta} e^{-f_\theta(\mathbf{x})} \right) \tag{13.80}$$

$$= \nabla_\mathbf{x} \log \frac{1}{Z_\theta} + \nabla_\mathbf{x} \log e^{-f_\theta(\mathbf{x})} \tag{13.81}$$

$$= -\nabla_\mathbf{x} f_\theta(\mathbf{x}) \tag{13.82}$$

$$\approx \mathbf{s}_\theta(\mathbf{x}), \tag{13.83}$$

where $\mathbf{s}_\theta(\mathbf{x})$ denotes a parametric approximation called score function which can be learned by a neural network by minimizing

$$\mathbb{E}_{p(\mathbf{x})} \left[\|\mathbf{s}_\theta(\mathbf{x}) - \nabla_\mathbf{x} \log p(\mathbf{x})\|^2 \right]. \tag{13.84}$$

There are various methods to train a score matching neural network without learning the data distribution $\log p(\mathbf{x})$ first. For our purposes, we mention the denoising score matching (Vincent, 2011) that operates by using a pair of clean and corrupted samples $(\mathbf{x}, \tilde{\mathbf{x}})$ with joint density $p_\sigma(\mathbf{x}, \tilde{\mathbf{x}}) = p_\sigma(\tilde{\mathbf{x}}|\mathbf{x})p(\mathbf{x})$. Then, the score matching objective is altered to match the log-conditional distribution instead

$$\mathbb{E}_{p(\mathbf{x}, \tilde{\mathbf{x}})} \left[\|\mathbf{s}_\theta(\mathbf{x}) - \nabla_{\tilde{\mathbf{x}}} \log p_\sigma(\tilde{\mathbf{x}}|\mathbf{x})\|^2 \right]. \tag{13.85}$$

Considering that the corruption follows as a Gaussian distribution, we have

$$\nabla_{\tilde{\mathbf{x}}} \log p_\sigma(\tilde{\mathbf{x}}|\mathbf{x}) = \frac{1}{\sigma^2}(\mathbf{x} - \tilde{\mathbf{x}}), \tag{13.86}$$

which corresponds to moving back from the noisy to clean sample, and thus learning this direction by the score network. The averaging over the joint distribution of clean and corrupted samples can be shown to be equivalent to the explicit score matching goal in (13.84). Detailed proof of this equivalence can be found in the appendix of the work of Vincent (2011).

13.2.8 *Langevin sampling*

After training, the score matching model is able to produce an approximation of the gradient of the probability, such that $s_\theta(\mathbf{x}) \approx \nabla_{\mathbf{x}} \log p(\mathbf{x})$. Therefore, we could use this model to generate data using the learned score function as a gradient in a random walk. The score function thus defines a vector field over the entire space of the data $\mathbf{x}$, pointing toward the local maxima. By learning the score function of the data distribution, we can generate samples by starting at any arbitrary point in the data space and iteratively following the score until the closest probability maximum is reached. This sampling procedure is known as Langevin dynamics and is mathematically described as

$$\mathbf{x}_{i+1} \leftarrow \mathbf{x}_i + c\nabla \log p(\mathbf{x}_i) + \sqrt{2c}\boldsymbol{\epsilon}, \quad i = 0, 1, \ldots, K, \qquad (13.87)$$

where $\mathbf{x}_0$ is randomly sampled from a prior distribution (such as uniform) and $\boldsymbol{\epsilon} \sim \mathcal{N}(\boldsymbol{\epsilon}; \mathbf{0}, \mathbf{I})$ is an extra noise term to ensure that the generated samples do not always collapse onto a the local maximum but random move around it for sampling diversity. Furthermore, because the learned score function is deterministic, the noise avoids falling into deterministic trajectories, which is particularly important when the initial position lies between multiple probability modes.

There are several problems with the standard score matching (Song and Ermon, 2019). When $\mathbf{x}$ lies on a low-dimensional manifold in a high-dimensional space, all points that are not on the low-dimensional manifold would have probability zero and the score function will be ill-defined. Moreover, the estimated score function trained via standard score matching is not accurate in low density regions because the model might not be learning from such low probability examples and will require many more iterations to converge on an accurate output. Lastly, Langevin dynamics sampling may not be correct when the distribution comprises of a a weighted mix of several sources since the log operation separates the mixing coefficient from the probabilities and then the gradient operation zeros the coefficients. In such cases, Langevin dynamics starting from a point between several modes in a Mixture of Gaussians will have an equal chance of arriving at each mode, despite the mode having different weights.

These three drawbacks are addressed by adding multiple levels of Gaussian noise to the data. This is done by choosing a sequence of noise levels $\{\sigma_t\}_{t=1}^{T}$ that define a sequence of progressively perturbed data distributions:

$$p_{\sigma_t}(\mathbf{x}_t) = \int p(\mathbf{x})\mathcal{N}(\mathbf{x}_t; \mathbf{x}, \sigma_t^2\mathbf{I})d\mathbf{x}. \tag{13.88}$$

Then, a neural network $\mathbf{s}_\theta(\mathbf{x}, t)$ is learned using score matching to learn the score function for all noise levels simultaneously:

$$\underset{\boldsymbol{\theta}}{\arg\min} \sum_{t=1}^{T} \lambda(t)\mathbb{E}_{p_{\sigma_t}(\mathbf{x}_t)} \left[\|\mathbf{s}_\theta(\mathbf{x}, t) - \nabla \log p_{\sigma_t}(\mathbf{x}_t)\|_2^2\right], \tag{13.89}$$

where $\lambda(t)$ is a positive weighting function that conditions on noise level t.

With this objective, the score matching at multiple noise levels exactly matches the objective in the variational diffusion model that was described before. Proving this equivalence is beyond the scope of the book. Interested reader should consult the original paper (Song and Ermon, 2019), where the authors propose annealed Langevin dynamics procedure in which samples are produced by running Langevin dynamics for each $t = T, T - 1, \ldots, 2, 1$ in sequence, with each subsequent sampling starting from the final samples of the previous simulation. The authors show that since the noise levels steadily decrease over time steps t, and the step size is reduced over time, the samples eventually converge into a true mode. This procedure is directly analogous to the Markovian process of the variational diffusion model.

13.2.9 *Mathematical comparison of DDIM, DDPM, and score matching*

Here we provide a comparison of the three diffusion models and estimation methods.

Score matching: Score matching was historically the first of the three approaches to be developed. The primary goal of score matching is to estimate the score of the data distribution, $\nabla_x \log p_{\text{data}}(x)$, without explicitly modeling the data distribution itself:

- **Model:** A score network, $s_\theta(x) : \mathbb{R}^D \to \mathbb{R}^D$, is used to approximate the score function.
- **Estimation:** Score matching minimizes the following objective function: $E_{p_{\text{data}}(x)}[\text{tr}(\nabla_x s_\theta(x)) + \frac{1}{2}\|s_\theta(x)\|_2^2]$. Minimizing this objective yields a score network that approximates the true score function.
- **Sampling:** Score matching, by itself, does not provide a sampling method. However, the estimated score function can be used in Langevin dynamics to generate samples from the data distribution.
- **Challenges:**

  – *Manifold hypothesis*: If the data lie on a low-dimensional manifold, the score is undefined, leading to inaccurate score estimation.
  – *Low data density regions*: Score estimation is challenging in areas of low data density due to the lack of training data in those regions.

DDPM: DDPMs were developed after score-matching methods. They address the challenges of score matching by introducing a forward diffusion process that gradually adds noise to the data, ensuring that the perturbed data distribution has full support over the data space:

- **Model:** A set of functions $\epsilon_\theta^{(t)}(x_t) : \mathbb{R}^D \to \mathbb{R}^D$ is learned for each timestep t in the forward diffusion process.
- **Estimation:** A variational lower bound to the log likelihood is maximized. A surrogate objective, $L_\gamma(\epsilon_\theta)$, is used for optimization:

$L_\gamma(\epsilon_\theta) := \sum_{t=1}^{T} \gamma_t E_{x_0 \sim q(x_0), \epsilon_t \sim N(0,I)}[\|\epsilon_\theta^{(t)}(\sqrt{\alpha_t}x_0 + \sqrt{1-\alpha_t}\epsilon_t) - \epsilon_t\|_2^2]$. This objective is equivalent to the denoising score matching objective for specific choices of the weighting coefficients γ.

- **Sampling:** The generative process approximates the reverse of the forward diffusion process, progressively denoising a noisy observation x_T sampled from the prior distribution.
- **Challenges:**
  - *Sampling efficiency:* Generating a single sample requires simulating a Markov chain for many steps, making sampling computationally expensive.

DDIM: DDIMs were developed as an improvement upon DDPMs. They leverage the same training objective as DDPMs but introduce a deterministic generative process:

- **Model:** The same set of functions $\epsilon_\theta^{(t)}(x_t)$ is learned as in DDPM.
- **Estimation:** The same objective function, $L_\gamma(\epsilon_\theta)$, is used for training as in DDPM. This results in a shared model between the two approaches.
- **Sampling:** DDIM utilizes a non-Markovian forward process and a deterministic reverse process. The sampling equation is $x_{t-1} = \sqrt{\alpha_{t-1}}(\frac{x_t - \sqrt{1-\alpha_t}\epsilon_\theta^{(t)}(x_t)}{\sqrt{\alpha_t}}) + \sqrt{1 - \alpha_{t-1} - \sigma_t^2} \cdot \epsilon_\theta^{(t)}(x_t) + \sigma_t\epsilon_t$, where σ_t controls the stochasticity. When $\sigma_t = 0$ for all t, the sampling becomes deterministic.
- **Advantages over DDPM:**
  - *Sample efficiency:* DDIM achieves similar sample quality to DDPM in significantly fewer steps.
  - *Sample consistency:* DDIM generates samples with consistent high-level features from the same initial latent variable, regardless of the sampling trajectory.
  - *Interpolation:* DDIM permits semantically meaningful interpolation in the latent space.
  - *Reconstruction:* DDIM enables encoding data samples to latent representations and decoding them back with low error.
- **Relationship to Langevin dynamics:** DDPM and score-based generative models employ Langevin dynamics for sampling, which relies on the score function. DDIM can be seen as a deterministic

modification of Langevin dynamics achieved by setting $\sigma_t = 0$ in its sampling equation. Annealed Langevin dynamics, a technique used in score-based models to improve sampling by gradually decreasing noise levels, could potentially be applied to DDIM as well.

13.3   Conditioning and Guidance of Diffusion Models

Using the score-based formulation of a diffusion model, conditional diffusion problem is formulated as learning problem of estimating $\nabla_{\mathbf{x}_t} \log p(\mathbf{x}_t|y)$, where y is the conditioning variable and the score of the conditional model is learned for all noise levels t. We omit the subscript $\mathbf{x}_t$ in the future Applying the Bayes rule, we have

$$\nabla \log p(\mathbf{x}_t|y) = \nabla \log \left(\frac{p(\mathbf{x}_t)p(y|\mathbf{x}_t)}{p(y)} \right) \tag{13.90}$$

$$= \nabla \log p(\mathbf{x}_t) + \nabla \log p(y|\mathbf{x}_t) - \nabla \log p(y) \tag{13.91}$$

$$= \underbrace{\nabla \log p(\mathbf{x}_t)}_{\text{unconditional score}} + \underbrace{\nabla \log p(y|\mathbf{x}_t)}_{\text{conditioning gradient}} \tag{13.92}$$

with the gradient of $\log p(y)$ with respect to $\mathbf{x}_t$ equal zero. The posterior $p(y|\mathbf{x}_t)$ can be viewed as a classifier, predicting the label y from the noisy data $\mathbf{x}_t$. The difficulty in training a classifier that would be robust enough to different levels of noise, and at the same time produce significant enough gradients to small perturbations of the input, makes this approach difficult to implement.

An alternative approach, known as "Classifier-Free" method uses the same score function formulation with the following "trick". First, we rewrite the classifier guidance with a weighting term γ as follows:

$$\nabla \log p(\mathbf{x}_t|y) = \nabla \log p(\mathbf{x}_t) + \gamma \nabla \log p(y|\mathbf{x}_t). \tag{13.93}$$

When $\gamma = 0$, the conditional diffusion model ignores the conditioning information, while for large γ, the conditional diffusion model strongly follows to the conditioning information. To derive the score function under Classifier-Free Guidance, we rearrange the posterior conditioning probability as

$$\nabla \log p(y|\mathbf{x}_t) = \nabla \log p(\mathbf{x}_t|y) - \nabla \log p(\mathbf{x}_t). \tag{13.94}$$

Substituting this into the conditional diffusion guidance equation we get a modified gradient expression

$$\nabla^{CFG} \log p(\mathbf{x}_t|y)$$

$$= \nabla \log p(\mathbf{x}_t) + \gamma \left(\nabla \log p(\mathbf{x}_t|y) - \nabla \log p(\mathbf{x}_t) \right) \tag{13.95}$$

$$= \nabla \log p(\mathbf{x}_t) + \gamma \nabla \log p(\mathbf{x}_t|y) - \gamma \nabla \log p(\mathbf{x}_t) \tag{13.96}$$

$$= \underbrace{\gamma \nabla \log p(\mathbf{x}_t|y)}_{\text{conditional score}} + \underbrace{(1 - \gamma) \nabla \log p(\mathbf{x}_t)}_{\text{unconditional score}}. \tag{13.97}$$

This allows avoiding having a classifier and simply use two training procedures — one for unconditional model and one for conditional model. The guidance regime is determined by the strength of the γ parameter. Rewriting this in terms of the noise matching procedure, one gets

$$\nabla \log p(y|\mathbf{x}_t) \approx \epsilon_\theta(\mathbf{x}_t; t|y) - \epsilon_\theta(\mathbf{x}_t; t), \tag{13.98}$$

giving the guidance equation

$$\epsilon_\theta^{\text{CFG}}(\mathbf{x}_t; t|y) = (1 + \gamma)\epsilon_\theta(\mathbf{x}_t; t|y) - \gamma\epsilon_\theta(\mathbf{x}_t; t)). \tag{13.99}$$

The conditioning operation can be implemented using various schemes, such as concatenation, multiplication, feature-wise linear modulation (FiLM), attention, and more, applied the network internal features, such as the upsampling steps in U-Net of the the noise prediction network. Moreover, the type and representation of the conditioning variable determine the ability to control the diffusion model from text, images, or other media input.

13.3.1 *Text prompts in stable diffusion and AudioLDM*

Contrastive learning has been used to create a joint embedding of different types of media in one shared space, notably Contrastive Language and Image Pretraining (CLIP) for text and images (Radford *et al.*, 2021), and Contrastive Langauge and Audio Pretraining (CLAP) for text and music (Wu *et al.*, 2022). Stable Diffusion uses the CLIP text encoder to transform text prompts into a sequence of tokens. These tokens represent the semantic content of the prompt in a latent space. CLIP (Contrastive Language-Image Pretraining) is a neural network model trained on a massive dataset of image — text pairs. It learns to associate images with their corresponding textual descriptions. In Stable Diffusion, a pretrained CLIP text encoder is used to transform text prompts into a sequence of tokens that capture the semantic meaning of the prompt. This encoded text information is then used to guide the diffusion model's image generation process. In a similar way, AudioLDM and MusicLDM use CLAP (Contrastive Language-Audio Pretraining) to produce general audio, Foley effect, and music from text prompt. The details of CLIP and CLAP are beyond the scope of this chapter, and the interested reader is referred to the above mentioned references for more details.

The encoded text tokens can be integrated into the diffusion model's denoising process through several different mechanism. In Stable Diffusion, a cross-attention mechanism is used to allow the model to selectively attend to relevant parts of the text prompt while generating the image. In AudioLDM, FiLM (Feature-wise Linear Modulation) is used to condition the LDMs on audio embeddings during training and on text embeddings during sampling. Both methods provide different ways of including conditioning information from the text embedding into the diffusion process.

In more detail, cross-attention allows the model to selectively attend to relevant parts of the conditioning information (text prompt in Stable Diffusion) at different stages of the generation process. This is done by calculating attention weights that determine the influence of each part of the conditioning information on the generated output as follows:

$$Q = \mathbf{W_Q}\phi(\mathbf{Z_t}),$$
$$K = \mathbf{W_K}\tau(\mathbf{E_{text}}),$$
$$V = \mathbf{W_V}\tau(\mathbf{E_{text}}), \tag{13.100}$$
$$\mathbf{Attention}(\mathbf{Q}, \mathbf{K}, \mathbf{V}) = \text{softmax}\left(\frac{\mathbf{QK}^\top}{\sqrt{d}}\right)\mathbf{V},$$

where $\mathbf{Z_t}$ represents the noisy feature map at time step t in the diffusion process, $\mathbf{E_{text}}$ represents the text embeddings obtained from a CLIP text encoder corresponding to the text prompt, $\phi(\cdot)$ is a learnable embedding function applied to the feature map, and $\tau(\cdot)$ is a learnable embedding function applied to the text embeddings. $\mathbf{W_Q}$, $\mathbf{W_K}$, $\mathbf{W_V}$ are the learnable projection matrices used to transform the embedded feature map and text embeddings into query ($\mathbf{Q}$), key ($\mathbf{K}$), and value ($\mathbf{V}$) matrices, respectively. d is the dimension of the query and key vectors.

Both the noisy feature map ($\mathbf{Z_t}$) from the diffusion process and the text embeddings ($\mathbf{E_{text}}$) from CLIP are transformed into a suitable latent space using learnable embedding functions ($\phi(\cdot)$ and $\tau(\cdot)$). The embedded feature map and text embeddings are projected into query, key, and value matrices using learned projection matrices. Next, the attention mechanism calculates the similarity between the query and key matrices using a scaled dot product, and the softmax function normalizes these similarities into attention weights. Finally, the attention weights are used to compute a weighted sum of the value matrix. This weighted sum represents the context vector, which captures the relevant information from the text prompt to guide the image generation process.

Alternatively, FiLM applies an affine transformation to the features of each layer in the Latent Diffusion Model (LDM),[2] scaling and shifting them based on the conditioning information (audio embedding during training, text embedding during sampling). The scaling and shifting parameters (γ and β) are predicted from the conditioning embedding. This allows the conditioning information to directly modulate the features at each layer, effectively controlling the information flow and guiding the generation process.

While both AudioLDM and Stable Diffusion employ U-Net architectures as their backbones, there are notable differences. AudioLDM uses convolutional layers in its U-Net to process the audio data, which is represented as mel-spectrograms. Stable Diffusion uses convolutional layers for the initial and final stages of its U-Net and Vision Transformers (ViTs) in the intermediate layers to process image data. ViTs are particularly well-suited for capturing long-range dependencies in image data, which is important for generating coherent images.

Another important aspect of Diffusion is the Continuous versus Discrete aspect of the latent space. Both Stable Diffusion and

---

[2]We further explain the idea of LDM versus DM and other difference in the architecture between Stable Diffusion and AudioLDM at the end of this paragraph.

AudioLDM employ a continuous latent space learned by a variational autoencoder (VAE). In AudioLDM, the VAE compresses the mel-spectrogram representation of the audio into this latent space, allowing for efficient modeling of the high-dimensional audio data. In Stable Diffusion, the VAE encodes images into this lower-dimensional latent space, which is then used by the U-Net for the denoising diffusion process.

In contrast to these two approaches, DiffSound (Yang *et al.*, 2023) employs a discrete latent space. It achieves discretization by using vector-quantized variational autoencoder (VQ-VAE) that maps the continuous latent representations to a finite set of discrete codebook entries. The distinction between continuous and discrete latent spaces has important implications for the capabilities and characteristics of the respective models. Continuous latent spaces offer more flexibility and allow for smooth interpolation and manipulation of the latent representations. This is beneficial for tasks like image editing, style transfer, or generating variations of existing data. Discrete latent spaces can be advantageous for tasks like compression or for modeling data with inherently discrete structures, such as language. However, the quantization might lead to a loss of information or introduce artifacts in the generated output.

13.3.2 *Beyond words: Guiding diffusion with non-textual cues*

While text prompts have revolutionized creative AI, diffusion models are increasingly embracing non-textual conditions, unlocking new levels of control and creativity. Two prominent examples of this trend are ControlNet and Adapters.

ControlNet (Zhang *et al.*, 2023) extends pretrained text-to-image models like Stable Diffusion, enabling guidance from various spatial inputs such as edges, depth maps, segmentation maps, and even human poses. Instead of directly modifying the massive pretrained model, ControlNet introduces a parallel, trainable copy of its encoding layers, connected via "zero convolutions". This approach protects the pretrained weights while allowing the ControlNet to learn how to inject specific spatial conditions into the generation process.

Adapters (Poth *et al.*, 2023), originating from Natural Language Processing, offer a versatile and efficient approach to specializing pretrained models. Small, modular layers are inserted within the model's architecture, enabling adaptation to specific tasks without extensive retraining. In the context of diffusion models, adapters are particularly well suited for handling non-textual conditions. For instance, T2I-Adapters can be trained to align a pre-trained text-to-image model with external control signals, like sketches or depth maps. This allows for fine-grained control over color, structure, and composition, empowering users to precisely shape the generated output.

Mathematical summary of the two method is as follows:

ControlNet: A ControlNet block takes an input feature map x and outputs a modified feature map y_c based on the original neural network block $F(\cdot; \Theta)$ and a trainable copy of the block with parameters Θ_c. The two copies are connected with zero convolution layers $Z(\cdot; \cdot)$:

$$y_c = F(x; \Theta) + Z(F(x + Z(c_f; \Theta_{z1}); \Theta_c); \Theta_{z2}),$$

where y_c is the output of the ControlNet block, x is the input feature map, Θ_{z1} and Θ_{z2} are parameters of the zero convolution layers, and c_f is the conditioning vector, obtained by encoding the input conditioning image c_i using a convolutional neural network $E(\cdot)$: $c_f = E(c_i)$. See Figure 13.5 for a diagram of the ControlNet architecture. The original neural network block $F(\cdot; \Theta)$ transforms an input feature map x into another feature map y:

$$y = F(x; \Theta),$$

where x and y are usually 2D feature maps. In Stable Diffusion (SD), the ControlNet block copies the Encoder and the Bottleneck (middle block) of the U-Net architecture.

Diffusion Model Learning Objective (Loss Function) is same as the standard diffusion model learning objective, which involves predicting the noise added to a noisy image at time step t

$$L = \mathbb{E}_{z_0, t, c_t, c_f, \epsilon \sim N(0,1)} \left[\|\epsilon - \epsilon_\theta(z_t, t, c_t, c_f)) \|_2^2 \right],$$

where L is the loss function, z_0 is the original input image, t is the time step representing the level of noise, c_t is the text prompt, c_f is the encoded conditioning image, ϵ is the added noise, and ϵ_θ is the

$$c_f = E(c_i)$$

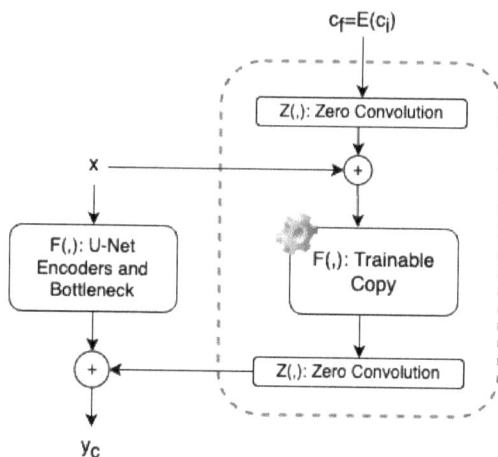

Fig. 13.5. ControlNet architecture.

noise prediction network. In the training process, 50% of the text prompts c_t are randomly replaced with empty strings to encourage the ControlNet to directly recognize semantic content in the input conditioning image.

Zero convolution is central to ControlNet as it bridges the original frozen model with the trainable copy at the condition input and the network output. It is characterized by having both its weights and biases initialized to zero. This initialization is crucial for ControlNet's functionality as it ensures that the zero convolution layers have no effect on the computation at the start of training.

While a zero convolution layer can technically perform scalar multiplication when the input has a single channel, since it is a convolutional layer, it can process multi-channel input feature maps, performing a weighted sum of the input channels at each spatial location. Moreover, because the weights are initialized to zero, this weighted sum initially outputs zero for all locations and channels. During training, the parameters of the zero convolution layers (Θ_{z1} and Θ_{z2}) are gradually adjusted. This gradual learning is important as it prevents the sudden introduction of potentially disruptive gradients into the frozen pretrained model. One may say that the zero convolution layers act as a buffer, protecting the frozen model from "harmful noise" during the initial training steps. This characteristic also contributes to the "sudden convergence phenomenon", as described in the original paper.

One may note that this is distinct from FiLM that uses a feature-wise affine transformation to modulate feature maps based on conditioning information. While both zero convolution and FiLM involve linear transformations, there are key differences. FiLM learns the parameters (γ and β) of the affine transformation from the conditioning input directly. In contrast, the parameters of zero convolution layers are initialized to zero and gradually learned during training. Moreover, FiLM is often applied in conjunction with normalization layers, and it directly modulates the feature maps based on the conditioning information. Zero convolution layers, on the other hand, indirectly influence the feature maps by controlling the information flow between the original model and the trainable copy, as explained above.

Adapters: Adapters originated in natural language processing (NLP) as a method to efficiently adapt large language models (LLMs) to new tasks or domains without needing to fine-tune all model parameters. As such, they became a central technique in so-called Parameter Efficient Fine Tuning (PEFL) (Han *et al.*, 2024) methods. In diffusion models, T2I-Adapters are widely adopted as lightweight modules that align the pre-trained model internal representations with external non-textual control signals, such as sketch, depth map, color palette, or other guiding information, without modifying the original model's architecture. Denoting by $\mathbf{F^c} = F_{AD}(\mathbf{C})$ the multiscale condition features extracted by the T2I-Adapter operating on $\mathbf{C}$ condition input, these features are injected into the Diffusion Model at different U-Net layers $\hat{\mathbf{F}}_{\mathbf{i}}^{\mathbf{enc}} = \mathbf{F}_{\mathbf{i}}^{\mathbf{enc}} + \mathbf{F}_{\mathbf{i}}^{\mathbf{c}}, \quad i \in \{1, 2, 3, 4\}$, where $\hat{\mathbf{F}}_{\mathbf{i}}^{\mathbf{enc}}$ represents the modified encoder features at scale i after incorporating the condition features $\mathbf{F}_{\mathbf{i}}^{\mathbf{c}}$ with the original pretrained unmodified diffusion model encoder features $\mathbf{F}_{\mathbf{i}}^{\mathbf{enc}}$ at scale i.

The training process for T2I-Adapters involves minimizing the difference between predicted and actual noise added to the latent representation of the input image. This process utilizes triplets of original image, condition map, and text prompt, similar to the training of

(a)

(b)

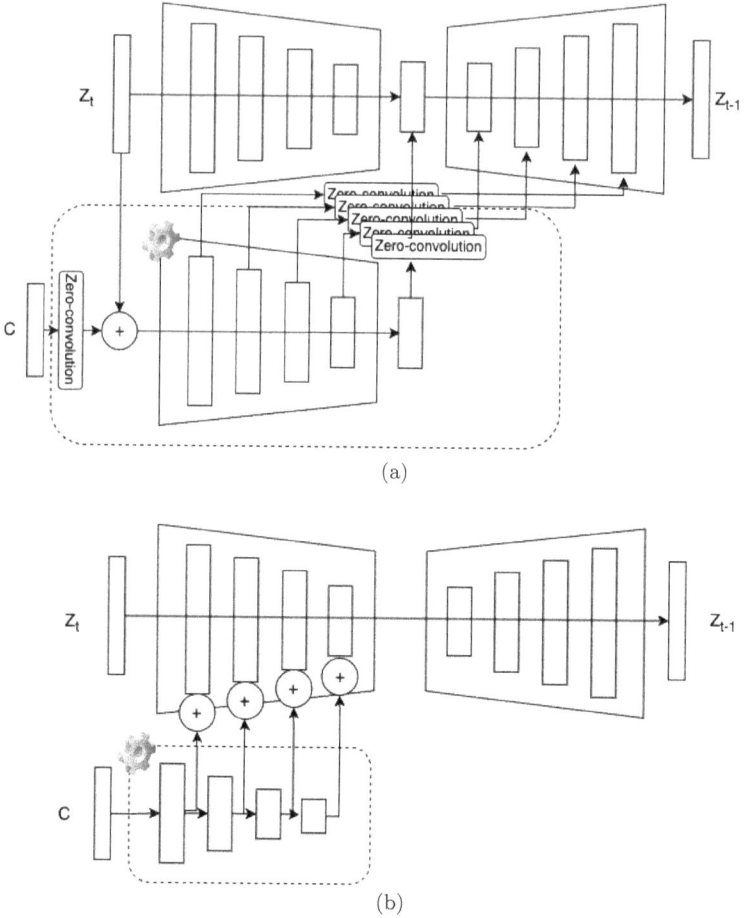

Fig. 13.6.   (a) ControlNet; (b) T2I adaptor.

Stable Diffusion. The parameters of the pretrained Stable Diffusion model are frozen during training, and only the T2I-Adapter parameters are updated. Figure 13.6 summarizes the two architectures.

In summary, ControlNet offers fine-grained control through iterative conditioning at each diffusion step, while T2I Adapters apply conditioning in a single pass during encoding, making the process more efficient but less adaptable during generation.

PART 5

Deeper Understanding of Deep Learning

Chapter 14

Information Theory of Learning

As Claude Shannon, the father of information theory, expressed in his work, *The Mathematical Theory of Communication* (Shannon 1949), the concept of information concerns the possibilities of a message rather than its content: 'That is, information is a measure of one's freedom of choice when one selects a message.' The measure of information is done by means of entropy, sometimes also called 'uncertainty'. The initial interest in information came from communications: how a message can be transmitted over a communication channel in the best possible way. The relevant questions in that setting were concerned compression and errors that happen during the encoding or transmission process. The models of information source were generative in the sense that it assumes an existence of a probability distribution from which the messages are randomly drawn.

14.1 Stochastic Modeling, Prediction, Compression, and Entropy

The underlying assumption in information-theoretical approach to machine learning is that a given data source can be produced by an unknown stochastic source. It statistics of the data can be learned by a machine, then production of new data by a generative model will affectively create new instances of the same information source. Same applies to compression — if a data can be efficiently compressed, then sampling and decoding from the compressed codes will create additional instances of data from the same source as well. This establishes a connection between compression and generation, or in case of temporal signals, between compression and prediction. This relation is formally expressed in terms of *asymptotic equipartition property* (AEP) (Cover and Thomas, 1991), which is the information theoretic analog to the law of large numbers in probability theory. The AEP tells us that if $x_1, x_2, \ldots$ are i.i.d random variables distributed with probability $P(x)$, then

$$-\frac{1}{n} \log_2 P(x_1, x_2, \ldots) \to H(P), \tag{14.1}$$

where $H(x)$ is the Shannon entropy of the $x \sim P(x)$, $H(x) = -\Sigma_x P(x) \log_2 P(x)$, where the averaging is over all possible occurrences of the sequences x.

The AEP property is graphically represented in Figure 14.1. The outer circle represents all possible sequences of length n, which are combinatorial number of possibilities depending on the size of the

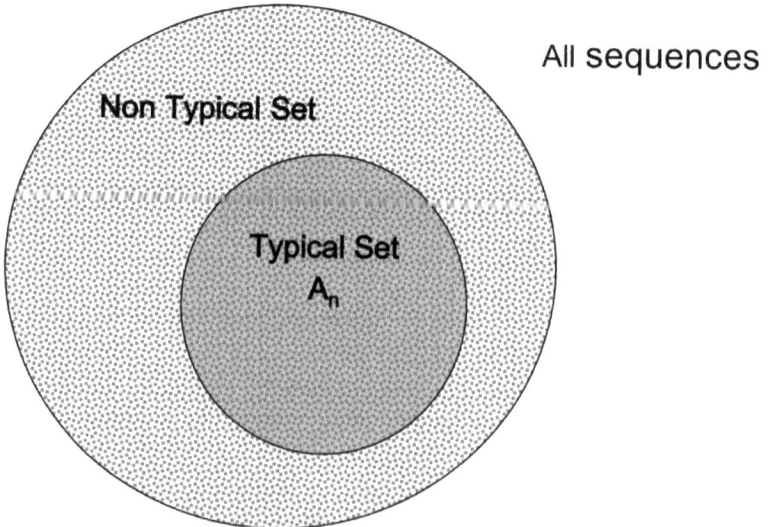

Fig. 14.1. Asymptotic equipartition property of long sequences.

alphabets. Shannon's theory proves that in view of the different probabilities of occurrence of each symbol (in the simplest case these are unbalanced heads or tails or Bernoulli distribution with binary choice), the entropy of the probability can be used to find a significantly smaller set of outcomes whose probability will tend to one, while other events tend to vanish. This set of outcomes is called the "Typical Set" and is denoted here as A_n, where n is the number of elements in a sequence that needs to be sufficiently large. Moreover, all sequences of the typical set are equiprobable, or in other words, one can index them in a way that no further structure or compression can be done. In our case, for generative purposes this means that we can access these strings using a uniform random number generator.

For stationary ergodic processes, and in particular, finite-order Markov processes, the generalization of AEP is called the Shannon–McMillan–Breiman theorem (Cover and Thomas, 1991). The connection with compression is that for long x the lower limit on compressibility is $H(x)$ bits per symbol. Thus, if we can find a good algorithm that reaches the entropy, then the dictionary of phrases it creates can be used for generating new instances from that source. Since the dictionary has to be very efficient, or in other words, it has to eliminate any other structure present in the sequence, or otherwise it could be compressed more, then we can now sample from the source by random selection from that dictionary. More specific aspects of how the dictionary is created and how continuity is being maintained between the random draws are discussed in the following.

In the case of Markov Chains, the amount if information can be quantified analytically as follows. Given a finite state space $S = s_1, \ldots, s_N$ and transition matrix $P(s_i, s_j) = a_{ij}$, which gives the probability of moving from a state i at some point in time t to another state j at the next step in time $t + 1$, under some mixing conditions, when a Markov process is started from any random initial state, after enough iterations a stationary distribution emerges. Denoting the stationary distribution as π, achieving a stationary state means that the probability of visiting any state remains unchanged for additional transition steps, which is mathematically expressed as $\pi = A\pi$, with A being the transitions matrix $A = [a_{ij}]$.

Entropy Rate of a random process is defined as the average entropy per symbol over a sequence of measurements, which is also equivalent of innovation entropy or the entropy of a random variable conditioned on its past. Given a sequence of random measurements $X^n = \{X_1, X_2, \ldots, X_n\}$ from a process $(X) = \{X_1, X_2, \ldots, X_n, \ldots\}$, the two definitions of entropy rate are

- Average Entropy per symbol

$$H_r(\mathcal{X}) = \lim_{n \to \infty} \frac{H(X^n)}{n}, \tag{14.2}$$

- Innovation Entropy

$$H'_r = \lim_{n \to \infty} H(X_n | X^{n-1}). \tag{14.3}$$

The two definitions are trivially equivalent when the events are independent since $H(X^n) = nH(X)$. For stationary process, since $H(X^n) = \sum_1^n H(X_i | X^{i-1})$, using the Cesaro mean

if $a_n \to a$ and $b_n = \frac{1}{n} \sum_{i=1}^n a_i$, then $b_n \to a$,

we get

$$H'_r(\mathcal{X}) \to H_r(\mathcal{X}). \tag{14.4}$$

Expressions for entropy of stationary distribution and the entropy rate in a Markov chain can be derived from the definition of conditional entropy:

$$H(\mathcal{X}) = \lim_{n \to \infty} H(X_n | X^{n-1}) = \lim_{n \to \infty} H(X_n | X_{n-1}) = H(X_2 | X_2). \tag{14.5}$$

Since by definition $P(X_2 = o_j | X_1 = s_i) = a_{ij}$, the entropy and the entropy rate are given by the following expressions:

$$H(S) = H(\pi) = -\Sigma_{i=1}^N \pi_i \log_2(\pi_i), \tag{14.6}$$

$$H_r(S) = H_r(A) = -\Sigma_{i=1}^N \pi_i \Sigma_{i=j}^N a_{ij} \log_2(a_{ij}). \tag{14.7}$$

Information rate (IR) combines the notions of entropy and entropy rate to explore how information propagates in time in a stochastic process. IR is defined as the difference between the entropy of the stationary distribution and the entropy rate in a stochastic process, which is equivalent to mutual information between past and future states. In the case of a Markov process, the past dependence is limited to current state, with IR given by

$$IR(S) = I(S_{t+1}, S_t) = H(\pi) - H_r(A). \tag{14.8}$$

For general stationary time series, this definition can be also extended to arbitrary past duration, viewed as information passing through time measured in terms of the mutual information between the present sample "as is" and its distribution when its past is taken into account. Denoting $\overleftarrow{X}_n = (\ldots, X_{n-2}, X_{n-1})$ as the *past* of X_n, and $\overrightarrow{X}_n = (X_{n+1}, X_{n+2}, \ldots)$, we may define several measure to capture the way information passes through time for a process with stationary (shift invariant) process (Abdallah and Plumbley, 2012):

- **Information rate**: $IR(X) = \rho_\mu = I(X_n, \overleftarrow{X}_n)$.
- **Predictive information rate**: $PIR(X) = b_\mu = I(X_n, \overleftarrow{X}_n), \overrightarrow{X}_n)$.

Taken together with the conditional entropy of the present sample given both its past and present $r_\mu = H(X_n | \overleftarrow{X}_n), \overrightarrow{X}_n)$, also called "erasure entropy", we can summarize the statistical relations between past, present, and future as follows in Figure 14.2.

Information Rate allows us to consider the advantage of making predictions, or in other word, the reduction in uncertainty that one might have when considering the next outcome in a random process by making a prediction versus considering only the statistics of the stationary distribution. Predictive information rate captures the difference in information rate between prediction of more then on step into the future and a single step entropy rate, as we show in the following.

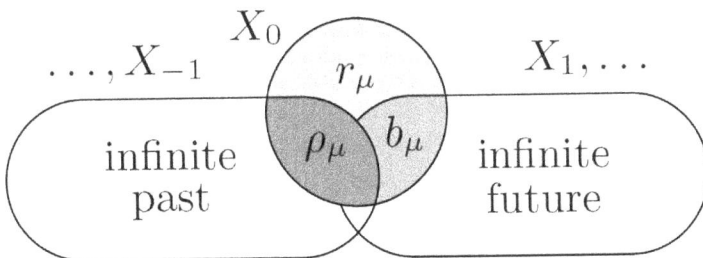

Fig. 14.2. The entropy of the present comprises of a sum of information rate, predictive information rate, and erasure entropy $H(X_n) = \rho_\mu + r_\mu + b_\mu$.

For example, if we consider a Markov chain that moves repeatedly from state 1 to N, with negligible probability for jumping between non-adjacent states. Such a situation can be described by matrix A that is nearly diagonal (the non-diagonal elements will be very small). One can verify that for such matrix, $H_r(A) \approx 0$ and IR will be close to the entropy of the stationary state IR $\approx \log_2(N)$. This is the maximal information rate for such process, showing that such very predictable process has high amount of dependency on its past. If on the contrary the matrix A is fully mixing with $a_{i,j} \approx \frac{1}{N}$, then the stationary distribution will be $\pi \approx \frac{1}{N}$ and the entropy and the entropy rate will be the same $H(S) = H_r(A) = \log_2(N)$. This results in IR close to zero, meaning that knowledge of the previous step tell close to nothing about the next step. Such process is very unpredictable.

Applying the notion of IR to Markov processes can guide the choice of Markov processes in terms of its predictive properties or average surprisal that such a process entails.

Predictive Information Rate measure for Markov chain is given as

$$\text{PIR}(S) = I(S_{t+1}, S_t | S_{t-1}) = I(S_{t+1}, (S_t, S_{t-1})) - I(S_{t+1}, S_{t-1}) \tag{14.9}$$

$$= H_r(A^2) - H_r(A).$$

Considering the previous example of a fully sequential repetitive process with $A \approx I$ diagonal matrix, one finds that since $H_r(A) \approx H_r(A^2)$, the resulting PIR is approximately zero. So, in terms of considering the mutual information for one step prediction given the past, since the process is nearly deterministic, knowing the past will determine both the next and the following steps, so there is little information passing from the present to the future when the past is known. For further analysis of these measures for discrete-time Gaussian processes, the reader is referred to the work of Abdallah and Plumbley (2012). Information Rate analysis can be extended to non-Markov and non-linear processes by considering models such as RNN, LSTM, time series use of CNNs, or Transformers. This requires measuring the entropy of the prediction error versus entropy of the process when knowledge of the temporal structure is absent, or in other words, when the samples are considered as i.i.d. What is common to these models is that the structure of the data is represented in terms of latent states, such as activations of hidden layers in neural networks. In following section, we introduce an information theoretical framework known as "Information Bottleneck" that tries to formalize the statistics of latent parameters in terms of information.

14.2 Information Bottleneck

The goal of Information Bottleneck is to propose a general framework for finding a hidden representation Z that can serve as an intermediary between two variables X and Y, in a way that will be both efficient and generalizable. The requirement on Z are that it efficiently captures the salient aspects of X which in turn can be used to predict Y. In this setting, the triplet X, Y, Z obeys Markov relation $p(X, Y, Z) = p(Y|X)p(X|Z)p(Z)$. When applied to analysis of machine learning models, X can be considered as an input to a neural network, Y are the output variables, with Z being the intermediate representation learned by tuning the parameters of the machine learning system, such as weights of the different layers in a multi-layer perceptron.

Using the Markov relations $Z - X - Y$, one can formulate the goal of a machine learning system as one of finding an optimal trade-off between simplicity of representation and its prediction ability. Such learning is looking for a representation that minimizes the discrepancy, or statistical difference, between signal prediction using complete information about the input, which is the case of time series comprised of the complete the past X, versus its prediction capability when using a simplified encoding of the input, or some representation if its the past Z. This error, averaged over all possible encoding pairs X, Z is given by

$$\langle D_{\mathrm{KL}}(p(Y|X)\|p(Y|Z)) \rangle_{p(X,Z)} = I(X, Y|Z) = I(X, Y) - I(Z, Y),$$

(14.10)

where $D_{\mathrm{KL}}(\cdot, \cdot)$ is the Kullback–Leibler (KL) divergence between two distributions and $I(\cdot, \cdot)$ is the mutual information between their random variables.

Since $I(X, Y)$ are independent of Z, minimizing the KL divergence happens when $I(Z, Y)$ is maximized, with zero KL obtained when $I(Z, Y) = I(X, Y)$.

Since minimizing KL could be trivially satisfied by taking $Z = X$, a constraint on Z needs to be added, such as requiring it to be the most compact or simplest latent "explanation" of X. In information theoretical terms, we can write this requirement as minimization of $I(X, Z)$. Combining the two goals, we arrive at the target function for our learning system:

$$\max_{P(Z|X)} \{I(Z, Y) - \lambda I(X, Z)\}. \tag{14.11}$$

The above expression is known as a Information Bottleneck (IB) (**?**). Our formulation shows that IB can be formulated as a KL minimization problem with a compression constraint on the laten representation.

Proof. Proof of the latent KL-IB relations (Dubnov, 2022).

In our notation, $D_{\mathrm{KL}}(\cdot, \cdot)$ is the Kullback–Leibler distance between different distributions and $I(\cdot, \cdot)$ is the mutual information between their random variables:

$$D_{\mathrm{KL}}(P, Q) = \int p(x) \log \frac{p(X)}{q(X)} dX \tag{14.12}$$

with Mutual information defined as

$$I(X, Y) = H(X) - H(X|Y) = H(Y) - H(Y|X)$$
$$= H(X) + H(Y) - H(X, Y) \tag{14.13}$$

and signal entropy given by

$$H(X) = -\int p(X) \log p(X) dX. \tag{14.14}$$

Another useful relation is between KL distance and Mutual Information, which can be derived from the above definitions:

$$I(X, Y) = D_{\mathrm{KL}}(p(X, Y), p(X)p(Y))$$
$$= D_{\mathrm{KL}}(p(X|Y)p(Y), p(X)p(Y)). \tag{14.15}$$

In other words, Mutual Information $I(X, Y)$ measures the KL distance between a joint distribution $P(X, Y) = p(X, Y)$ and marginal distribution $Q(X, Y) = p(X)p(Y)$. Taking into account the Markov relations $Z - X - Y$, we have the following conditional independence relations between our variables $p(Y, X, Z) = p(Y|X, Z)p(X, Z) = p(Y|X)p(X, Z)$. Using the definition of KL divergence and the averaging over $P(X, Z)$, we have

$$\langle D_{\text{KL}}(p(Y|X)\|p(Y|Z)) \rangle_{p(X,Z)}$$

$$= \int p(X, Z) \left(\int p(Y|X) \log \frac{p(Y|X)}{p(Y|Z)} dY \right) dX dZ$$

$$= \int p(X, Z, Y) \log \left(\frac{p(Y|X)p(X)}{p(X)p(Y)} \frac{p(Y)p(Z)}{p(Y|Z)p(Z)} \right) dY dX dZ$$

$$= \int p(X, Z, Y) \log(\frac{p(Y, X)}{p(X)p(Y)} - \int p(X, Z, Y) \log(\frac{p(Y)p(Z)}{p(Y, Z)}$$

$$= I(X, Y) - I(Z, Y),$$

where we used the KL and Mutual information relations shown above.

The last remaining relation we need to show is that $I(X, Y|Z) = I(X, Y) - I(Z, Y)$. This can be proved by considering the definition of mutual information as

$$I(X, Y) = H(Y) - H(Y|X),$$

$$I(Z, Y) = H(Y) - H(Y|Z)$$

and using the Markov relation $H(Y|X, Z) = H(Y|X)$ to see that $I(X, Y|Z) = H(Y|Z) - H(Y|X, Z) = H(Y|Z) - H(Y|X) = H(Y) - H(Y|X) - H(Y) + H(Y|Z) = I(X, Y) - I(Z, Y)$.

Since the mutual information between X and Y is independent of Z, we can remove $I(X, Y)$ and set the KL minimization goal as a maximization of $I(Z, Y)$. To avoid a trivial solution of the Z minimization problem by $Z = Y$, we add a constraint of minimization of $I(X, Z)$.

Combining the two goals, the target function for the learning system becomes the maximization over $P(Z|X)$ of a combined expression

$$\max_{P(Z|X)} \{ I(Z, Y) - \lambda I(X, Z) \}. \tag{14.16}$$

$\square$

In other words, one may postulate that a goal of a machine learning system is finding a latent representation Z that "explains out" most of the mutual information between input and output of a neural network. In the case of time-series modeling, the IB applies to modeling of the stochastic process by finding a model whose latent parameters preserve the maximal Information Dynamics of the data $I(X, Y)$ with X being the past and Y the next sample prediction. This principle is expressed as minimization of $I(X, Y | Z)$, i.e., finding a latent reduced Z, so that there will be minimal erasure or residual information passing between the past X and the present Y in the temporal signal. In a such case, the IB principle is called "predictive IB".

IB principle says that a goal of a learning system is to find the most compact representation Z of X that still provides most information about a different variable Y. Accordingly, predictive IB looks for the most compact representation of the past that carries maximal information about the time series future. Additionally, the fidelity of representation of the signal X by its reduced representation Z has to be accounted. This is done by adding a distortion $D(X, Z)$ that captures the reconstruction quality of X when decoding it from Z.

To summarize, one may formulate the following criteria that combine competing goals of three learning factors $I(X, Z)$, $I(Y, Z)$, and $d(X, Z)$:

- finding the most compact representation of present X that is most informative about the Y (in the case of temporal signals, this is the information dynamics interpretation) and
- finding the most compact representation of X from which X can be recovered with minimal distortion $d(X, Z)$ (i.e., reconstruction quality).

We identify the first criteria as predictive information and the second as representation learning.

14.3 Bits-Back Coding in VAE

Ordinary coding of data maps each input instance to a unique codeword. According to Shannon's theory, the best encoding of a data is lower bounded by its entropy, so finding the best code requires knowing the correct data distribution in order to assign code words of appropriate length to each data instance. In latent models, the distribution of data x depends on the value of a latent variable z. When we consider z as side information that can be transmitted in parallel with the data itself, then the ambiguity of x in terms of its hidden "meaning" that is captured by z can be used efficiently in the decoding process.

Bits-back coding is casting the problem of optimizing VAE into a problem of compression by designing an efficient coding scheme in a situation where the data probabilities depend on an additional auxiliary variable z. To do that, VAE is seen as a way to encode data into two main parts: $p(z)$ that encodes the prior/side knowledge about the essential latent or hidden structures of the data, and $p(x|z)$ being the realization (decoding) of the data from it's latent representation. The variational approximation adds an additional auxiliary part that transmits the "meaning" of the data (inference of z, or finding the hidden cause) as determined by the encoder $q(z|x)$.

In such a scenario, the transmission of x can be done in two steps: encoding of the true latent variable z first and then encoding of x using the knowledge of z. The expected code length under this coding scheme is given by a sum of the two codes:

$$C(x) = E_{x \sim p(x), z \sim q(z|x)}[-\log p(z) - \log p(x|z)]$$
$$= H(x, z) = H(x) + H(z|x),$$

where $1/\log p(z)$ and $1/\log p(x|z)$ are the Shannon code lengths of z, and x given z, respectively. This adds an additional code penalty $H(z|x)$ to the ideal efficient encoding of the data x in terms of Shannon theory $H(x)$. So, why would we want to infer and possibly transmit or share knowledge about abstract data structures between sender and receiver if this amounts to wasting computational and communication resources? The solution to this paradox comes from the bits-back argument that remediates the extra coding length penalty by making the inference of z available at the decoder. In such a case, z does not need to be transmitted, as it can be inferred from x alone, reducing the coding $C(x)$ back to $H(x)$:

$$C_{\text{BitsBack}}(x) = \mathbb{E}_{x \sim p(x), z \sim q(z|x)}[-\log p(z) - \log p(x|z)]$$
$$= \mathbb{E}_{x \sim p(x)}[-\log p(x) + D_{\text{KL}}(q(z|x)\|p(z|x))]$$
$$= H(x) + \mathbb{E}_{x \sim p(x)}[D_{\text{KL}}(q(z|x)\|p(z|x))].$$

Only when $q(z|x) = p(z|x)$, i.e., the listener has perfect knowledge of the latent causes, does the extra coding penalty become zero. It can be said that by randomly selecting codewords according to probability distribution $q(z|x)$ in the decoder, we in fact communicate extra auxiliary information along with the data we are coding. The bits-back method provides an encoding scheme, details of which are not proved here, that claims that the extra information communicated through an auxiliary data is in fact a knowledge that is shared between the encoder and decoder.

The interested reader is referred to Frey and Hinton's Free Energy Coding articles (Frey and Hinton, 1996) where they propose a distribution approximation that uses the joint coding length of x and z as an energy function. In the paper, they work out a toy example of a Gaussian mixture model and show that the encoding of data by always encoding it in terms of a single maximum-likelihood Gaussian and sending the Gaussian index as side information performs worse than using an ambiguous coding that randomly selects the Gaussians at the decoder. This example is shown in Figure 14.3. So, instead of transmitting the side information and doing what seems to be the most efficient encoding for each sample, it is actually better to embrace the ambiguity about the causes (i.e., which Gaussian most likely generated the data) since the identity of the Gaussian that

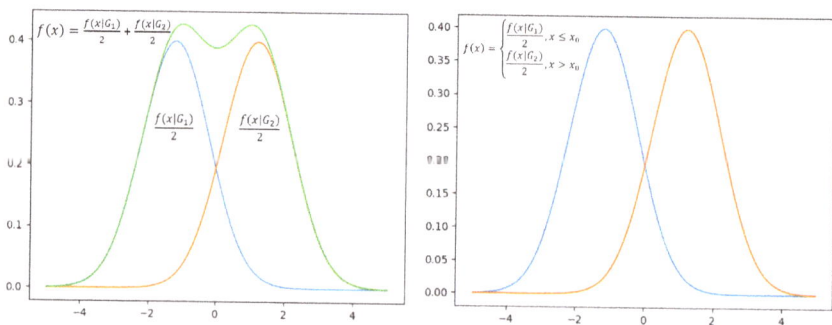

Fig. 14.3. This example shows a source model that may produce multiple codewords for a given symbol. The graph on left shows a source with a single binary hidden variable which identifies from which Gaussian, G_1 or G_2, the symbol value x is sampled. Values of x near x_o are likely to have come from either Gaussian. The graph on right shows the resulting coding density effectively used if we were to always pick the shorter codeword. This density wastes coding space because it is wrongly shaped.

produces a given symbol is unknown, especially in regions where the Gaussians strongly overlap and the data is likely to have come from either Gaussian. In these cases, the source model maps each symbol to two codewords, one for each Gaussian, and is left up to the decoder to randomly decide how the code is interpreted. It turns out that by randomly selecting codewords, the method effectively communicates the extra side information about the hidden source along with the data it is coding. Looking at this differently, the bits communicated in the auxiliary side information will make up for the excess codeword lengths that result from not always using the shortest codeword, thus getting the bits back.

The encoding scheme of the bits-back argument operates when a source code produces multiple codewords, or in other words, when ambiguity exists as to origin of the data with respect to possible origin or cause of the message. In this scheme, the auxiliary data provided in the encoder are recovered by the receiver. By randomly selecting codewords, the bits communicated in the auxiliary data will make up for the excess codeword lengths that result from not always using the shortest codeword.

This approach was used as a tool for discovering efficient perceptual codes (Frey *et al.*, 1997). In information theoretical terms, these codes are communicated between a sender and a receiver.

An illustration, modified from the original paper, is provided in Figure 14.4 for the case of musical communication. If the music signal are well modeled in terms of musical features like chroma

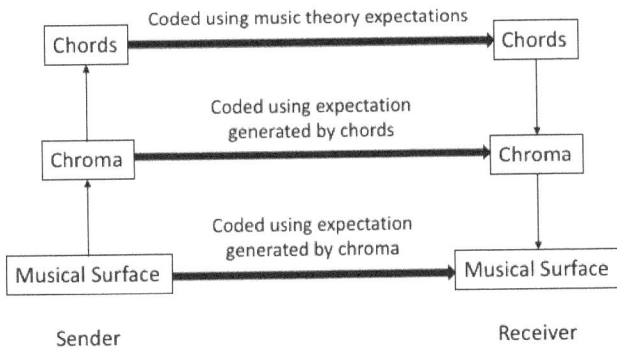

Fig. 14.4. An illustration of the relationship between good representations and economical communication from the work of Frey and Hinton (1996).

that represents the total energy present in the different tones, and chords that represent musically valid common combinations of tones according to music theory, and lastly, rules of chord progressions and substitutions, then the musical "surface" that represents the actual musical data, can be communicated more economically (or meaningfully for the musician or the listener) by using this representation. The top-down expectations that are produced by the generative model operating on representations at one level will assign high probabilities to the data produced at the next level down, and so on, so by using the top-down expectations, it is possible to communicate the data efficiently.

For our purposes, since we are not interested in compression *per se*, we want to use the bits-back argument to understand the tradeoffs between encoding of $p(x|z)$ with or without substantial influence of z. As shown in (14.17), the two-part code view of VAE adds to the the minimal representation of the data $H(x)$ an extra code length of $D_{\mathrm{KL}}(q(z|x)\|p(z|x))$ nats for using a posterior that is not precise. We may understand now why sometimes the latent code z is not appropriately used by considering it at as an inefficiency of VAE as stated by the bits-back coding: A long sequence of input data $x_1, x_2, \ldots, x_n$ that is modeled by decoding from distribution $p(x_1, x_2, \ldots, x_n|z)$ without access to z will be captured locally in terms of structural relations that govern the dynamics of the time series regardless of it's latent representation z. The "remedy" to the situation of VAE ignoring z can be two fold — either encoding small portions of data that do not require very powerful decoders, thus breaking the representation into chunks or fine small granularity where the relations between z and x are strongly manifested, or encoding larger data structures in approximate ways, such as using lossy encoding that prefers global statistics and discards local statistics. When we learn a lossy compression/representation of data, we can try to construct a decoding distribution just for reconstructing the detailed parts of the data that we don't want the lossy representation to capture. In other words, the lossy encoding should be done in a way that the this encoder is incapable of modeling the detailed information that the lossy representation discards. This trade-off between lossy latent encoding efficiency versus representation error is discussed in the following.

14.4 Can IB Explain the Dynamics of Deep Leaning?

So far, we have seen several interpretations of representation learning as lossy compression that finds an optimal trade-off between encoding complexity (codes that maximally compress the input) and quality of the decoding (codes that are most informative or predictive with respect to the target variable). The question we want to discuss here is whether this phenomenon holds for general neural networks, and even more specifically, we would like to know of IB emerges during the learning process so that it can serve as a fundamental principle that explains the dynamics of the learning process in deep neural network (DNN) architectures.

To do so, one may consider a general MLP architecture in terms of a Markov chain that maps an input X to output Y through multiple hidden layers $h_1, h_2, \ldots, h_m$, where the output of the activation function of each layer is considered as a random variable. Using mutual information between every layer h_i and the network input and output variables, it is possible to visualize the learning dynamics of the network in *information plane* whose axes are $I(X, h_i), I(h_i; Y))$, with each layer represented as a point in that plane.

In the information plane, each representation variable, h_i, is considered as a stochastic map of the input X. The statistical relation between the data (input) X and the labels (output) Y is characterized by its joint distribution $P(X, Y)$. The relation to the representation variable is similarly characterized by its encoder and decoder distributions, $P(X, h_i)$ and $P(h_i, Y)$, respectively. For a DNN of K-layers, the layers are mapped to K monotonically connected points in the plane — henceforth a unique information path — which satisfies the following DPI chains:

$$I(X; Y) \geq I(h_1; Y) \geq I(h_2; Y) \geq \cdots \geq I(h_k; Y) \geq I(\hat{Y}; Y),$$
$$(14.17)$$

$$H(X) \geq I(X; h_1) \geq I(X; h_2) \geq \cdots \geq I(X; h_k) \geq I(X, \hat{Y}).$$
$$(14.18)$$

These inequalities represent a general trend of information relations between the training data variables X, Y and the hidden network parameters, comprising a graph of points the monotonically decreases with layer number. It should be noted that the last layer (the network output) is denoted as $\hat{Y}$, while in the information plane analysis, the two axes are the data variables X, Y themselves. Accordingly, that first layers appear higher in the upper right corner in the information plane, with all subsequent layers laying lower and to the left.

The analysis of the learning dynamics comprises of tracing the information plan position of the different layers at every time step of the training process. This visualization gives an intuitive representation of the dynamics of neural network learning in terms of its information content. In the work of Tishby and Zaslavsky (2015), it was demonstrated that the process of *Stochastic Gradient Descent* for a neural network tends to decompose into two stages:

- Fitting the network to the data ("moving the funnel"). During this stage, $I(h_i, Y)$ and $I(h_i, X)$ grow.
- Compressing the output of the network ("shrinking the funnel"). During this stage, $I(h_i, Y)$ grows slower and $I(h_i, X)$ shrinks, which is seen as reversing the direction of the layer on the $I(h_i, X)$ axis, representing increase in compression of the input (stronger encoding), while the decoding continues to improve.

To get further insight into the dynamics of learning, we consider whether the learning process achieves an optimal IB solution. If we define Z as the compressed representations of X, the IB problem is formulated as an optimization over for the distributions, $p(Z|X), p(Z), p(Y|Z)$, with the assumption that Z is independent of Y given X, or in other words, that they constitute a Markov chain $Y - X - Z$:

$$\min_{p(Z|X), p(Z), p(Y|Z)} \{I(X, Z) - \beta I(Z, Y)\}. \qquad (14.19)$$

The general solution to the problem is given by Blahut–Arimoto algorithm, which we bring here without a proof. An interested reader is referred to Chapter 10 of Cover and Thomas book on information theory (?). The IB case is solved in a manner similar to the Rate-Distortion case, replacing the distortion function with $D_{\text{KL}}(p(y|x)\|p(y|z))$.

14.5   Blahut–Arimoto Algorithm for IB

Given a source signal X with probability $p(X)$, we find an encoding of X into a compressed signal Z that minimizes the mutual information $I(X, Z)$ through a conditional probability $p(Z|X)$ by repeating the following iteration until convergence:

$$p_{t+1}(Z) = \sum_X p(X)p_t(Z|X), \tag{14.20}$$

$$p_{t+1}(Z|X) = \frac{p_t(Z)\exp(-\beta I(X, Z))}{\sum_Z p_t(Z)\exp(-\beta D_{\mathrm{KL}}(p(Y|X)\|p(Y|Z)))}. \tag{14.21}$$

An example of the IB optimal curve is given in Figure 14.5 for an example of a simple Markov chain model knows as "Noisy Type-writer" that comprises of an alphabet of 64 input letters (thus 6 bits)

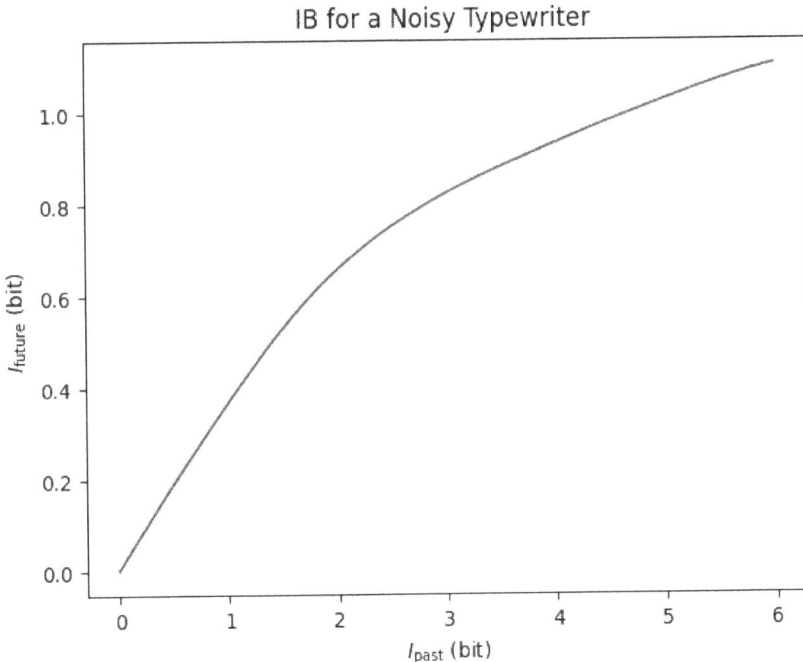

Fig. 14.5.   An example of IB achievable compression for an example of a Noisy Typewriter.

where each letter is randomly mistyped into an equally probable choice among its left and right 16 neighboring letters. Although this is not a DNN model, we bring it here for the purpose of demonstrating the achievable optimal compression for a noisy channel between random variables X and Y. The simulation was conducted using the EMBO toolbox (Piasini and Gold, 2021). For $P(X,Y)$ distributions, the information curve is strictly concave with a slope, β^{-1}, representing the compression amount at every point on the curve. It should be noted that for deterministic networks, one may consider the sigmoidal output of the neurons as probabilities, which is the common in using cross-entropy and log-loss error in the stochastic optimization.

Considering the optimality of IB, there is growing evidence of results that demonstrate that in many cases the dynamics of DNN learning follow the IB dynamics, with final hidden layer parameters settling close to the optimal IB curve. This evidence lead to a claim, initially made in the work of Tishby and Zaslavsky (2015), that compression phase is important for the generalization ability of DNNS. Contrary to this claim, in the work of Saxe *et al.* (2018), the authors have shown that compression dynamics in the information plane does not always occur, strongly depending on the nonlinearities employed by the network. Furthermore, it was shown that generalization performance does not always follow the information plane, thus putting the compression-generalization hypothesis under question. Finding out what type of DNN actually achieves IB optimal bounds still remains an open research question. Nevertheless, the information bottleneck principle seems to provide an important insight and a principled theoretical framework for further study of deep networks.

Chapter 15

NN as Gaussian Processes

15.1 Gaussian Processes

In the previous chapters, we have seen the tremendous power of neural networks and their applications in various areas. We also have briefly analyzed the approximation powers of fully connected neural networks in Chapter 2. In this chapter, we look at neural networks from a different view: We regard a neural network as a random function where the source of randomness are the parameters. At the end of this chapter, we reach the conclusion that when a neural network is very wide, it acts as a Gaussian process, or a Gaussian random field.

We first review the concept of Gaussian processes. A more comprehensive introduction can be found in, e.g., the work of Williams and Rasmussen (2006). A Gaussian process (GP) is a probability distribution over functions $y(\mathbf{x})$ defined on $\mathbb{R}^D$ such that for an arbitrary collection of points $(\mathbf{x}_1, \ldots, \mathbf{x}_N)$, the joint distribution of $y(\mathbf{x}_1), \ldots, y(\mathbf{x}_N)$ is Gaussian. When $D > 1$, it is also custom to use the name *Gaussian random field*.

Obviously, from the definition, if $y(\mathbf{x})$ is a GP, then for any fixed $\mathbf{x}$, $y(\mathbf{x})$ follows a Gaussian distribution. Note that a GP may or may not be stationary: $y(\mathbf{x})$ may or may not follow the same distribution for different positions of $\mathbf{x}$. An example of non-stationary GP is a Brownian motion. We illustrate 1,000 realizations of a Brownian motion in Figure 15.1. Consider the horizontal axis t as time axis and the vertical axis y as space axis. Clearly, from the paths, for a larger time t, the Gaussian $y(t)$ has a larger variance.

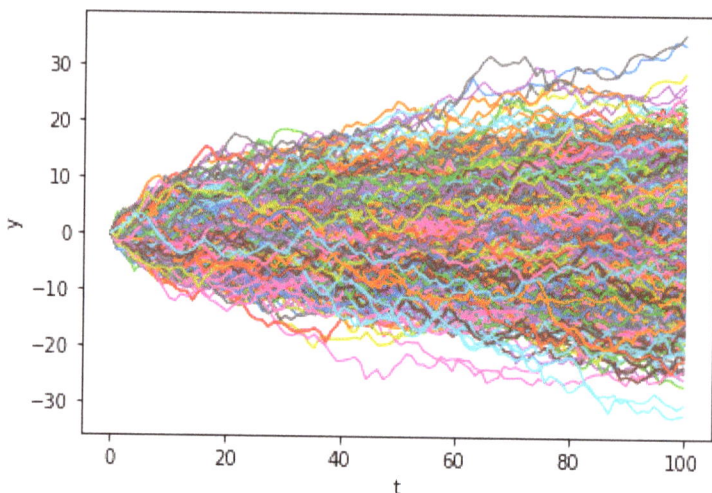

Fig. 15.1. Thousand realizations of a Brownian motion.

A GP is completely determined by two functions: The first is the expectation function $\mu(\mathbf{x}) := \mathbb{E}[y(\mathbf{x})]$, and the second is the covariance function $K(\mathbf{x}, \mathbf{x}') := \text{Cov}(y(\mathbf{x}), y(\mathbf{x}'))$. The function K is also called the kernel function. We thus denote a GP as $\mathcal{GP}(\mu, K)$. In machine learning, a GP can be used to describe our prior belief. In applications where we do not have prior knowledge about the mean of $y(\mathbf{x})$, it is custom to set $\mu(\mathbf{x}) = 0$ by symmetry. On the other hand, there are commonly used kernel functions as prior choices. For instance, the linear kernel satisfies $K(\mathbf{x}, \mathbf{x}') = \mathbf{x}^\top \mathbf{x}'$, the radial basis function (RBF) kernel satisfies

$$K(\mathbf{x}, \mathbf{x}') = \exp\left(-\frac{\|\mathbf{x} - \mathbf{x}'\|^2}{2s^2}\right),$$

and the Ornstein–Uhlenbeck kernel satisfies

$$K(\mathbf{x}, \mathbf{x}') = \exp\left(-\frac{\|\mathbf{x} - \mathbf{x}'\|}{s}\right).$$

Figure 15.2 illustrates realizations of GPs with two different kernels.

GP lends itself to important applications in regression problems. Consider a linear kernel regression model $y(\mathbf{x}) = \mathbf{w}^\top \phi(\mathbf{x})$, where $\mathbf{x}$ is the input, ϕ is a pre-defined nonlinear transformation, and $\mathbf{w}$ is

(a)

(b)

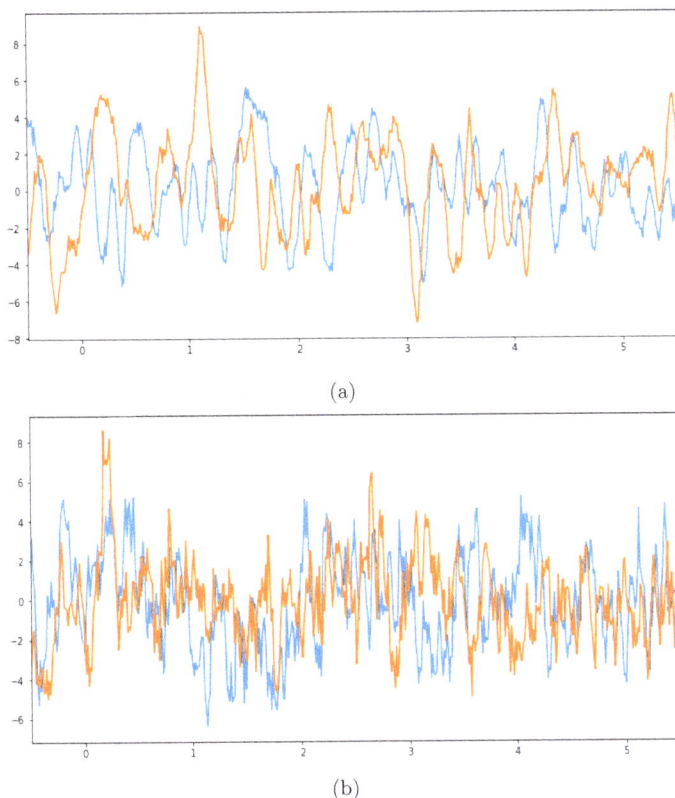

Fig. 15.2. Illustration of two GPs with different kernels. Each GP has two different realizations (a) with an RBF kernel. (b) with an Ornstein–Uhlenbeck kernel.

the vector of weights. Suppose we impose a very simple prior on $\mathbf{w}$: $p(\mathbf{w}) = \mathcal{N}(\mathbf{x}|\mathbf{0}, \mathbf{I})$. Then, $y(\mathbf{x})$ is $\mathcal{GP}(\mu, K)$, where $\mu(\mathbf{x}) = 0$ and

$$K(\mathbf{x}, \mathbf{x}') = \phi(\mathbf{x})^\top \phi(\mathbf{x}'). \tag{15.1}$$

That is, in the case of kernel linear regression, imposing a prior on $\mathbf{w}$ is equivalent to imposing the GP prior on $y(\mathbf{x})$. We emphasize again that we don't need to consider a ϕ in practice and should directly go with a proper choice of K.

Consider the regression model where the received label is generated by $t = y + \epsilon$, where $\epsilon \sim \mathcal{N}(0, \sigma^2)$. Equivalently, we have $p(t|y) = \mathcal{N}(t|y, \sigma^2)$. Consider the collection of training data with N points whose input features are $\mathbf{x}_1, \ldots, \mathbf{x}_N$ and the corresponding labels are assembled in $\mathbf{t}_N = (t_1, \ldots, t_N)^\top$, conditioned on $\mathbf{y}_N = (y_1, \ldots, y_N)^\top$. We have

$$p(\mathbf{t}_N | \mathbf{y}_N) = \mathcal{N}(\mathbf{t}_N | \mathbf{y}_N, \sigma^2 \mathbf{I}). \tag{15.2}$$

Also, according to the GP assumption, the joint prior is given by

$$p(\mathbf{y}_N) = \mathcal{N}(\mathbf{y}_N | \mathbf{0}, \mathbf{K}), \tag{15.3}$$

where $\mathbf{K}$ is an $N \times N$ matrix whose (n, m)th entry is given by $\mathbf{K}(n, m) = K(\mathbf{x}_n, \mathbf{x}_m)$. Combining (15.2) and (15.3), the marginal distribution is given by

$$p(\mathbf{t}_N) = \mathcal{N}(\mathbf{t}_N | \mathbf{0}, \mathbf{L}_N), \tag{15.4}$$

where $\mathbf{L}_N$ is an $N \times N$ matrix given by

$$\mathbf{L}_N = \mathbf{K}_N + \sigma^2 \mathbf{I}. \tag{15.5}$$

Now, consider a test point $\mathbf{x}_{\text{test}}$. We would like to predict the corresponding label by finding $p(t_{\text{test}} | \mathbf{t}_N)$. We proceed by first considering the joint distribution $p(\mathbf{t}_{N+1})$, where $\mathbf{t}_{N+1} = (t_1, \ldots, t_N, t_{\text{test}})^\top$. Following the same way of deriving (15.4), we have

$$p(\mathbf{t}_{N+1}) = \mathcal{N}(\mathbf{t}_{N+1} | \mathbf{0}, \mathbf{L}_{N+1}), \tag{15.6}$$

where $\mathbf{L}_{N+1}$ is an $(N + 1) \times (N + 1)$ matrix given by

$$\mathbf{L}_{N+1} = \mathbf{K}_{N+1} + \sigma^2 \mathbf{I}. \tag{15.7}$$

Here, $\mathbf{K}_{N+1}$ can be "extended" from $\mathbf{K}_N$ by

$$\mathbf{K}_{N+1} = \begin{bmatrix} \mathbf{K}_N & \mathbf{k}_N \\ \mathbf{k}_N^\top & k_{N+1} \end{bmatrix}, \tag{15.8}$$

where the nth entry of $\mathbf{k}_N$ is $K(\mathbf{x}_n, \mathbf{x}_{\text{test}})$ and $k_{N+1} = K(\mathbf{x}_{\text{test}}, \mathbf{x}_{\text{test}})$. Following Gaussian analysis, the joint Gaussian implies that the condition density $p(t_{\text{test}} | \mathbf{t}_N)$ is also Gaussian $\mathcal{N}(\mu_{\text{test}}, \sigma^2_{\text{test}})$, whose mean is given by

$$\mu_{\text{test}} = \mathbf{k}_N^\top \mathbf{L}_N^{-1} \mathbf{t}_N \tag{15.9}$$

and whose variance is given by

$$\sigma_{\text{test}}^2 = k_{N+1} - \mathbf{k}_N^\top \mathbf{L}_N^{-1} \mathbf{k}_N. \tag{15.10}$$

15.2 NN-GP for a Simple Fully Connected Network

A neural network is a function $y(\mathbf{x}; \boldsymbol{\theta})$ determined by its parameters $\boldsymbol{\theta}$. When $\boldsymbol{\theta}$ carries the randomness, $y(\mathbf{x})$ can be thought of as a random process. When the neural network is very wide, i.e., the number of neurons in each layer is very large, the corresponding random process is a GP.

Let's first consider the simplest case where we have a fully connected network. Let $\mathbf{x} \in \mathbb{R}^{d_0}$ be the input. Let $\tilde{\mathbf{y}} = f^{(1)}(\mathbf{x}) = \mathbf{W}^{(1)}\mathbf{x}$ be the linear transformation for the first layer. For simplicity, let's omit the bias terms since including them does not make a difference in principle. The hidden layer represents $\mathbf{y} = g^{(1)}(\mathbf{x}) = \sigma(f^{(1)}(\mathbf{x})) \in \mathbb{R}^{d_1}$, where σ is a nonlinear activation function. The output is given by $\mathbf{z} = f^{(2)}(\mathbf{y}) = \mathbf{W}^{(2)}\mathbf{y} \in \mathbb{R}^{d_2}$. To facilitate discussion, we write the matrix-vector multiplications explicitly in indices:

$$y_j = \sigma\left(\sum_{k=1}^{d_0} W_{jk}^{(1)} x_k\right), \quad j = 1, \dots, d_1, \tag{15.11}$$

$$z_i = \sum_{j=1}^{d_1} W_{ij}^{(2)} y_j, \quad i = 1, \dots, d_2. \tag{15.12}$$

Figure 15.3 illustrates this neural network.

Now, we need a specific initialization for the weights of the above neural network. We take $W_{jk}^{(1)} \sim \mathcal{N}(0, \sigma_W^2)$ and $W_{ij}^{(2)} \sim \mathcal{N}(0, \sigma_W^2/d_1)$ for each i, j, k where the parameters are independent. Note that this is one of many possible settings that facilitate use of easy notations. First, we consider $\tilde{y}_j = \sum_{k=1}^{d_0} W_{jk}^{(1)} x_k$. As a linear combination of Gaussian variables, $\tilde{y}_j$ itself is also Gaussian. Moreover, when considered as a function of $\mathbf{x}$, $\tilde{y}_j(\mathbf{x})$ is clearly a GP since $\{\tilde{y}_j(\mathbf{x})\}_{\mathbf{x} \in \mathbb{X}}$ is jointly Gaussian for any finite collection $\mathbb{X}$ of input features. Specifically,

$$\mathbb{E}[\tilde{y}_j(\mathbf{x})] = \mathbb{E}\left[\sum_{k=1}^{d_0} W_{jk}^{(1)} x_k\right] = \sum_{k=1}^{d_0} 0 \cdot x_k = 0 \tag{15.13}$$

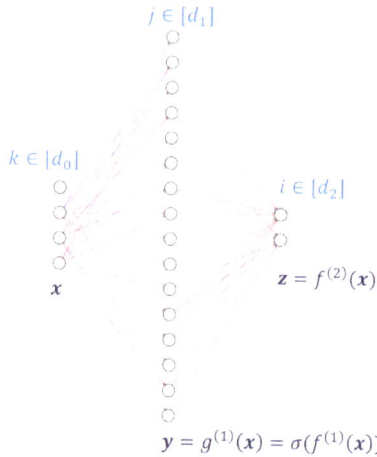

Fig. 15.3.   Illustration of an NN with a single hidden layer.

and

$$\text{Cov}(\tilde{y}_j(\mathbf{x}), \tilde{y}_j(\mathbf{x}')) = \mathbb{E}[\tilde{y}_j(\mathbf{x})\tilde{y}_j(\mathbf{x}')] \tag{15.14}$$

$$= \mathbb{E}\left[\sum_{k=1}^{d_0} \sum_{k'=1}^{d_0} W_{jk}^{(1)} x_k W_{jk'}^{(1)} x_{k'}'\right] \tag{15.15}$$

$$= \mathbb{E}\left[\sum_{k=1}^{d_0} \left(W_{jk}^{(1)}\right)^2 x_k x_k'\right] \tag{15.16}$$

$$= \sigma_W^2 \sum_{k=1}^{d_0} x_k x_k' = \sigma_W^2 \mathbf{x}^\top \mathbf{x}'. \tag{15.17}$$

Next, we consider (15.12). We define $\tilde{W}_{ij}^{(2)} = \sqrt{d_1} W_{ij}^{(2)}$ for each i, j. Then,

$$z_i = \frac{1}{\sqrt{d_1}} \sum_{j=1}^{d_1} \tilde{W}_{ij}^{(2)} y_j. \tag{15.18}$$

According to (15.11), y_j has an identical distribution for all $j = 1, \ldots, d_1$. Therefore, the variable $\tilde{W}_{ij}^{(2)} y_j$ also has an identical distribution, whose expectation is zero, for all j. By the Central Limit Theorem, as $d_1 \to \infty$, the distribution of z_i will tend to a centered

Gaussian distribution. It is also clear that $z_i(\mathbf{x})$, considered as a function of $\mathbf{x}$, will tend to a GP as $d_1 \to \infty$. The collection $\mathbf{z}$ at the end is also a GP where functions are vector-valued.

Let's consider z_i at infinite width ($d_1 \to \infty$) as a GP. Clearly, the mean of the GP is constantly zero. It is also not difficult to calculate the covariance. Since $\tilde{W}_{ij}^{(2)} \sim \mathcal{N}(0, \sigma_W^2)$,

$$\mathbb{E}[z_i(\mathbf{x})z_i(\mathbf{x}')] = \mathbb{E}\left[\frac{1}{d_1}\sum_{j=1}^{d_1}\tilde{W}_{ij}^{(2)}y_j(\mathbf{x})\sum_{j'=1}^{d_1}\tilde{W}_{ij'}^{(2)}y_{j'}(\mathbf{x}')\right] \quad (15.19)$$

$$= \mathbb{E}\left[\frac{1}{d_1}\sum_{j=1}^{d_1}\left(\tilde{W}_{ij}^{(2)}\right)^2 y_j(\mathbf{x})y_j(\mathbf{x}')\right] \quad (15.20)$$

$$= \sigma_W^2\,\mathbb{E}\left[y_1(\mathbf{x})y_1(\mathbf{x}')\right]. \quad (15.21)$$

In the last expression, y_1 can be replaced with any y_j since they are identically distributed. Note that we can apply our understanding of $\tilde{y}_j$ here. More specifically,

$$\mathbb{E}[y_1(\mathbf{x})y_1(\mathbf{x}')] = \mathbb{E}[\sigma(\tilde{y}_1(\mathbf{x}))\sigma(\tilde{y}_1(\mathbf{x}'))] \quad (15.22)$$

$$= \mathbb{E}_{(u,v)\sim\mathcal{N}(\mathbf{0},\mathbf{\Lambda}^{(1)})}[\sigma(u)\sigma(v)], \quad (15.23)$$

where

$$\mathbf{\Lambda}^{(1)} = \sigma_W^2\begin{bmatrix}\mathbf{x}^\top\mathbf{x} & \mathbf{x}^\top\mathbf{x}' \\ \mathbf{x}'^\top\mathbf{x} & \mathbf{x}'^\top\mathbf{x}'\end{bmatrix}. \quad (15.24)$$

Together, we have

$$\mathbb{E}[z_i(\mathbf{x})z_i(\mathbf{x}')] = \sigma_W^2\,\mathbb{E}_{(u,v)\sim\mathcal{N}(\mathbf{0},\mathbf{\Lambda}^{(1)})}[\sigma(u)\sigma(v)]. \quad (15.25)$$

Hence, we have fully expressed the GP that describes each z_i.

15.3   More General NN-GP

The analysis in the previous section can be generalized to NN with more than two layers. Note that the same results hold if we take parameters $W_{ij}^{(2)} \sim \mathcal{N}(0, \sigma_W^2)$ and take

$$z_i = \frac{1}{\sqrt{d_1}} \sum_{j=1}^{d_1} W_{ij}^{(2)} y_j \tag{15.26}$$

in place of (15.12). For easy representation in deeper layers, we follow the work of Arora *et al.* (2019b) and take $\sigma_W^2 = 1$. Moreover, let's consider a normalizing factor

$$c_\sigma = (\mathbb{E}_{z \sim \mathcal{N}(0,1)}[\sigma(z)]^2)^{-1}. \tag{15.27}$$

Now, we can recursively define a well-normalized fully connected network as follows. Suppose the NN contains a input layer with d_0 neurons, L hidden layers with $d_1, \ldots, d_L$ neurons, respectively, and a single output. We denote the input feature as

$$g^{(0)}(\mathbf{x}) = \mathbf{x} \in \mathbb{R}^{d_0}. \tag{15.28}$$

Then, for $l = 1, \ldots, L$,

$$f^{(l)}(\mathbf{x}) = \mathbf{W}^{(l)} g^{(l-1)}(\mathbf{x}) \in \mathbb{R}^{d_l}, \tag{15.29}$$

$$g^{(l)}(\mathbf{x}) = \sqrt{\frac{c_\sigma}{d_l}} \sigma(f^{(l)}(\mathbf{x})) \in \mathbb{R}^{d_l}, \tag{15.30}$$

where c_σ is defined as in (15.27). The output of the NN is

$$f(\mathbf{x}; \boldsymbol{\theta}) = f^{(L+1)}(\mathbf{x}) = \mathbf{W}^{(L+1)} g^{(L)}(\mathbf{x}) \in \mathbb{R}. \tag{15.31}$$

Here, $\boldsymbol{\theta}$ summarizes all the weights in $\mathbf{W}^{(1)}, \ldots, \mathbf{W}^{(L+1)}$. In a single equation,

$$f(\mathbf{x}; \boldsymbol{\theta}) = \mathbf{W}^{(L+1)} \sqrt{\frac{c_\sigma}{d_L}} \sigma \left(\mathbf{W}^{(L)} \cdots \sqrt{\frac{c_\sigma}{d_1}} \sigma(\mathbf{W}^{(1)} \mathbf{x}) \right). \tag{15.32}$$

Assume that all the weights are initialized as $W_{ij}^{(l)} \sim \mathcal{N}(0, 1)$ for all i, j, l. As $d_1, \ldots, d_L \to \infty$, for the same reason as discussed in the previous section, $f^{(1)}(\mathbf{x}), \ldots, f^{(L)}(\mathbf{x})$ and $f(\mathbf{x}; \boldsymbol{\theta})$ will tend to GPs whose mean is constantly zero. Now, let's calculate the covariance of $f(\mathbf{x}; \boldsymbol{\theta})$. First,

$$\mathbb{E}[f_i^{(1)}(\mathbf{x})f_i^{(1)}(\mathbf{x}')] = \mathbb{E}\left[\sum_{j=1}^{d_0} W_{ij}^{(1)}g_j^{(0)}(\mathbf{x}) \sum_{j'=1}^{d_0} W_{ij'}^{(1)}g_{j'}^{(0)}(\mathbf{x}')\right] \quad (15.33)$$

$$= \mathbb{E}\left[\sum_{j=1}^{d_0} g_j^{(0)}(\mathbf{x})g_j^{(0)}(\mathbf{x}')\right] \quad (15.34)$$

$$= \sum_{j=1}^{d_0} x_j x_j' = \mathbf{x}^\top \mathbf{x}' =: \Sigma^{(0)}(\mathbf{x}, \mathbf{x}'). \quad (15.35)$$

Next, for $l = 1, \ldots, L$, assume that we already have defined

$$\Sigma^{(l-1)}(\mathbf{x}, \mathbf{x}') := \mathbb{E}[f_i^{(l)}(\mathbf{x})f_i^{(l)}(\mathbf{x}')]. \quad (15.36)$$

We proceed for $\Sigma^{(l)}$ as follows:

$$\Sigma^{(l)}(\mathbf{x}, \mathbf{x}') = \mathbb{E}[f_i^{(l+1)}(\mathbf{x})f_i^{(l+1)}(\mathbf{x}')] \quad (15.37)$$

$$= \mathbb{E}\left[\sum_{j=1}^{d_l} W_{ij}^{(l+1)}g_j^{(l)}(\mathbf{x}) \sum_{j'=1}^{d_l} W_{ij'}^{(l+1)}g_{j'}^{(l)}(\mathbf{x}')\right] \quad (15.38)$$

$$= \mathbb{E}\left[\sum_{j=1}^{d_l} g_j^{(l)}(\mathbf{x})g_j^{(l)}(\mathbf{x}')\right] \quad (15.39)$$

$$= \mathbb{E}\left[\sum_{j=1}^{d_l} \sqrt{\frac{c_\sigma}{d_l}}\sigma\left(f_j^{(l)}(\mathbf{x})\right) \sqrt{\frac{c_\sigma}{d_l}}\sigma\left(f_j^{(l)}(\mathbf{x}')\right)\right] \quad (15.40)$$

$$= \frac{c_\sigma}{d_l}\sum_{j=1}^{d_l} \mathbb{E}\left[\sigma\left(f_j^{(l)}(\mathbf{x})\right)\sigma\left(f_j^{(l)}(\mathbf{x}')\right)\right] \quad (15.41)$$

$$= c_\sigma \mathbb{E}\left[\sigma\left(f_1^{(l)}(\mathbf{x})\right)\sigma\left(f_1^{(l)}(\mathbf{x}')\right)\right] \quad (15.42)$$

$$= c_\sigma \mathbb{E}_{(u,v)\sim\mathcal{N}(\mathbf{0},\boldsymbol{\Lambda}^{(l)})}[\sigma(u)\sigma(v)], \quad (15.43)$$

where

$$\mathbf{\Lambda}^{(l)} = \begin{bmatrix} \Sigma^{(l-1)}(\mathbf{x},\mathbf{x}) & \Sigma^{(l-1)}(\mathbf{x},\mathbf{x}') \\ \Sigma^{(l-1)}(\mathbf{x}',\mathbf{x}) & \Sigma^{(l-1)}(\mathbf{x}',\mathbf{x}') \end{bmatrix}. \tag{15.44}$$

That is, starting with (15.35), the recursive relations (15.44) and (15.43), $l = 1, \ldots, L$, determine $\mathbb{E}[f_i^{(L+1)}(\mathbf{x})f_i^{(L+1)}(\mathbf{x}')] = \mathbb{E}[f(\mathbf{x};\theta)f(\mathbf{x}';\theta)]$, which is exactly the covariance function of the GP. In short notation, we can denote the GP for $f(\mathbf{x};\theta)$ as $\mathcal{GP}(0,\Sigma^{(L)})$.

15.4 Neural Tangent Kernel

15.4.1 *Motivation*

In Chapter 15, we have seen that thanks to the Central Limit Theorem, infinitely wide neural networks can be regarded as Gaussian processes. We have also calculated the covariance, or the kernel function, of such GP. In this chapter, we look at a different kernel, called the *neural tangent kernel*, which is the correlation between the gradients of the output of NN with respect to the parameters, given a pair of input features. Let $f(\mathbf{x};\theta)$ denote the NN where $\mathbf{x}$ denotes the input feature and θ denotes the vector that summarizes the parameters of the NN. The neural tangent kernel is then

$$K(\mathbf{x},\mathbf{x}') = \left\langle \frac{\partial f(\mathbf{x};\theta)}{\partial\theta}, \frac{\partial f(\mathbf{x}';\theta)}{\partial\theta} \right\rangle. \tag{15.45}$$

First, let's intuitively understand why (15.45) is interesting. Consider a linear kernel regression model $f(\mathbf{x};\theta) = \theta^\top \phi(\mathbf{x})$, which is much simpler than a neural network. The corresponding kernel, used in kernel regression, is nothing but

$$K(\mathbf{x},\mathbf{x}') = \phi(\mathbf{x})^\top \phi(\mathbf{x}'). \tag{15.46}$$

However, in this case, because of the linear relation, we can represent $\phi(\mathbf{x})$ differently by

$$\phi(\mathbf{x}) = \frac{\partial f(\mathbf{x};\theta)}{\partial\theta}. \tag{15.47}$$

Then, clearly, (15.46) reduces to (15.45). In the case of linear kernel regression, the kernel is very important in representing the solution.

For instance, given a training set with input features $\{\mathbf{x}_n\}_{n=1}^N \subset \mathbb{R}^d$ and corresponding target output $\{y_n\}_{n=1}^N \subset \mathbb{R}$, the optimal linear kernel regression model is given by

$$f(\mathbf{x}) = \mathbf{k}(\mathbf{x})^\top \mathbf{K}^{-1} \mathbf{y}, \tag{15.48}$$

where $\mathbf{y} = (y_1, \ldots, y_N)^\top$, $\mathbf{k}(\mathbf{x}) = (K(\mathbf{x}, \mathbf{x}_1), \ldots, K(\mathbf{x}, \mathbf{x}_N))^\top$, and $\mathbf{K}$ is an $N \times N$ matrix whose (n, m)-th entry is given by $K(\mathbf{x}_n, \mathbf{x}_m)$. A detailed derivation of kernel regression can be found in the work of Bishop and Nasrabadi (2006, Chapter 6).

Another motivation for using the gradient is related to the gradient descent method for training neural networks. Consider again a training set $\{(\mathbf{x}_n, y_n)\}_{n=1}^N \subset \mathbb{R}^d \times \mathbb{R}$. Consider an NN represented by $f(\mathbf{x}; \boldsymbol{\theta})$ and a loss function

$$\ell(\boldsymbol{\theta}) = \frac{1}{2} \sum_{n=1}^N (f(\mathbf{x}_n; \boldsymbol{\theta}) - y_n)^2. \tag{15.49}$$

The gradient steps satisfy

$$\boldsymbol{\theta}^{(\tau+1)} = \boldsymbol{\theta}^{(\tau)} - \eta \nabla \ell(\boldsymbol{\theta}^{(\tau)}), \quad \tau = 0, 1, 2, \ldots. \tag{15.50}$$

Suppose we consider $\boldsymbol{\theta}$ as not a sequence but a function of continuous time $\boldsymbol{\theta}(t)$. We can set a schedule $\boldsymbol{\theta}^{(\tau)} = \boldsymbol{\theta}(\eta\tau)$. In this case, suppose the current time $t = \eta\tau$. We can reduce (15.50) to

$$\boldsymbol{\theta}(t + \eta) = \boldsymbol{\theta}(t) - \eta \nabla \ell(\boldsymbol{\theta}(t)), \tag{15.51}$$

or equivalently,

$$\frac{\boldsymbol{\theta}(t + \eta) - \boldsymbol{\theta}(t)}{\eta} = -\nabla \ell(\boldsymbol{\theta}(t)). \tag{15.52}$$

In the limit where $\eta \to 0$, that reads

$$\frac{d\boldsymbol{\theta}(t)}{dt} = -\nabla \ell(\boldsymbol{\theta}(t)). \tag{15.53}$$

Now, let's calculate the gradient on the right-hand side. Given the loss function defined in (15.49), we have

$$-\nabla \ell(\boldsymbol{\theta}(t)) = -\sum_{m=1}^N (f(\mathbf{x}_m; \boldsymbol{\theta}(t)) - y_m) \frac{\partial f(\mathbf{x}_m; \boldsymbol{\theta}(t))}{\partial \boldsymbol{\theta}}. \tag{15.54}$$

On the other hand, the derivative of the function value of the NN at $\mathbf{x}_n$, $n = 1, \ldots, N$, with respect to time is

$$\frac{df(\mathbf{x}_n; \boldsymbol{\theta}(t))}{dt} = \left\langle \frac{\partial f(\mathbf{x}_n; \boldsymbol{\theta}(t))}{\partial \boldsymbol{\theta}}, \frac{d\boldsymbol{\theta}(t)}{dt} \right\rangle \tag{15.55}$$

$$= -\sum_{m=1}^{N} (f(\mathbf{x}_m; \boldsymbol{\theta}(t)) - y_m)$$

$$\times \left\langle \frac{\partial f(\mathbf{x}_n; \boldsymbol{\theta}(t))}{\partial \boldsymbol{\theta}}, \frac{\partial f(\mathbf{x}_m; \boldsymbol{\theta}(t))}{\partial \boldsymbol{\theta}} \right\rangle. \tag{15.56}$$

Let $\mathbf{K}(t)$ denote the $N \times N$ matrix whose (n, m)th entry is given by

$$\mathbf{K}_{nm}(t) = \left\langle \frac{\partial f(\mathbf{x}_n; \boldsymbol{\theta}(t))}{\partial \boldsymbol{\theta}}, \frac{\partial f(\mathbf{x}_m; \boldsymbol{\theta}(t))}{\partial \boldsymbol{\theta}} \right\rangle. \tag{15.57}$$

We have

$$\frac{df(\mathbf{x}_n; \boldsymbol{\theta}(t))}{dt} = -\sum_{m=1}^{N} (f(\mathbf{x}_m; \boldsymbol{\theta}(t)) - y_m) \, \mathbf{K}_{nm}(t). \tag{15.58}$$

If we summarize the function values of the NN in a vector $\mathbf{u}(t) \in \mathbb{R}^N$ whose n-th entry is $f(\mathbf{x}_n; \boldsymbol{\theta}(t))$, then (15.58) can be written in the following vector form:

$$\frac{d\mathbf{u}(t)}{dt} = -\mathbf{K}(t)(\mathbf{u}(t) - \mathbf{y}). \tag{15.59}$$

Note that, according to the definition in (15.45), $\mathbf{K}$ is simply the kernel matrix for the data $\{\mathbf{x}_n\}_{n=1}^{N}$ for the neural tangent kernel. Therefore, the neural tangent kernel governs the dynamics of neural networks during training.

In general, $\mathbf{K}(t)$ is a function in time and changes with the NN training. Nevertheless, at infinite width, it has been shown that $\mathbf{K}(t)$ will tend to a fixed kernel $\mathbf{K}^*$ that does not vary with time. In this case, let $\mathbf{v}(t) = \mathbf{u}(t) - \mathbf{y}$. We have

$$\frac{d\mathbf{v}(t)}{dt} = -\mathbf{H}^* \mathbf{v}(t), \tag{15.60}$$

from which we can formally derive

$$\mathbf{v}(t) = e^{-t\mathbf{K}^*}\mathbf{v}(0), \tag{15.61}$$

where $e^{-t\mathbf{K}^*} = \sum_{j=0}^{\infty} \frac{(-t)^j}{j!}(\mathbf{K}^*)^j$. That is,

$$\mathbf{u}(t) = \mathbf{y} + e^{-t\mathbf{K}^*}(\mathbf{u}(0) - \mathbf{y}). \tag{15.62}$$

When $t \to \infty$, we validate that $\mathbf{u}(t) \to \mathbf{y}$.

15.4.2 *Derivation of the neural tangent kernel*

The proof that $\mathbf{K}(t) \to \mathbf{K}^*$ as the NN width approaches infinity is technical. Nevertheless, assuming this holds true, we can derive the exact form of $\mathbf{K}^*$. Our setting follows (15.28)–(15.30) in Chapter 15. We need to calculate

$$\mathbf{H}^*(\mathbf{x}, \mathbf{x}') = \left\langle \frac{\partial f(\mathbf{x}; \boldsymbol{\theta})}{\partial \boldsymbol{\theta}}, \frac{\partial f(\mathbf{x}'; \boldsymbol{\theta})}{\partial \boldsymbol{\theta}} \right\rangle. \tag{15.63}$$

Since $\mathbf{H}^*$ is constant over time, we can simply calculate it using the initially randomized parameters. The gradients are calculated in a backpropagation manner.

First of all, recall that

$$f(\mathbf{x}; \boldsymbol{\theta}) = \underbrace{\mathbf{W}^{(L+1)}}_{\in \mathbb{R}^{1 \times d_L}} \underbrace{\mathbf{g}^{(L)}(\mathbf{x})}_{\in \mathbb{R}^{d_L \times 1}}, \tag{15.64}$$

from which we have

$$\frac{\partial f(\mathbf{x}; \boldsymbol{\theta})}{\partial \mathbf{W}^{(L+1)}} = \left(g^{(L)}(\mathbf{x})\right)^{\top}. \tag{15.65}$$

Second of all, we write

$$f(\mathbf{x}; \boldsymbol{\theta}) = \mathbf{W}^{(L+1)} \underbrace{\sqrt{\frac{c_\sigma}{d_L}} \overbrace{\sigma(\mathbf{W}^{(L)} g^{(L-1)}(\mathbf{x}))}^{=f^{(L)}(\mathbf{x})}}_{=g^{(L)}(\mathbf{x})}, \tag{15.66}$$

from which we have

$$\frac{\partial f(\mathbf{x}; \boldsymbol{\theta})}{\partial \mathbf{W}^{(L)}} = \sum_{i=1}^{d_L} \frac{\partial f(\mathbf{x}; \boldsymbol{\theta})}{\partial g_i^{(L)}(\mathbf{x})} \frac{\partial g_i^{(L)}(\mathbf{x})}{\partial f_i^{(L)}(\mathbf{x})} \frac{\partial f_i^{(L)}(\mathbf{x})}{\partial \mathbf{W}^{(L)}}. \tag{15.67}$$

Clearly, for each $i = 1, \ldots, d_L$,

$$\frac{\partial f(\mathbf{x}; \boldsymbol{\theta})}{\partial g_i^{(L)}(\mathbf{x})} = \mathbf{W}_i^{(L+1)}, \tag{15.68}$$

where $\mathbf{W}_i^{(L+1)}$ is the ith entry (column) of $\mathbf{W}^{(L+1)}$;

$$\frac{\partial g_i^{(L)}(\mathbf{x})}{\partial f_i^{(L)}(\mathbf{x})} = \sqrt{\frac{c_\sigma}{d_L}} \dot{\sigma}\left(f_i^{(L)}(\mathbf{x})\right), \tag{15.69}$$

where $\dot{\sigma}$ is the derivative of the activation function σ; and

$$\frac{\partial f_i^{(L)}(\mathbf{x})}{\partial \mathbf{W}^{(L)}} = \begin{bmatrix} \mathbf{0} \\ \left(g^{(L-1)}(\mathbf{x})\right)^\top \\ \mathbf{0} \end{bmatrix}, \tag{15.70}$$

where $\left(g^{(L-1)}(\mathbf{x})\right)^\top$ is the ith row of the matrix. Altogether, we can represent $\partial f(\mathbf{x}; \boldsymbol{\theta})/\partial \mathbf{W}^{(L)}$ in a compact form as follows:

$$\frac{\partial f(\mathbf{x}; \boldsymbol{\theta})}{\partial \mathbf{W}^{(L)}} = \sqrt{\frac{c_\sigma}{d_L}} \overbrace{\underbrace{\mathbf{D}^{(L)}(\mathbf{x})}_{\in \mathbb{R}^{d_L \times d_L}} \underbrace{\left(\mathbf{W}^{(L+1)}\right)^\top}_{\in \mathbb{R}^{d_L \times 1}}}^{=:b^{(L)}(\mathbf{x})} \underbrace{\left(g^{(L-1)}(\mathbf{x})\right)^\top}_{\in \mathbb{R}^{1 \times d_L}}, \tag{15.71}$$

where $\mathbf{D}^{(L)}(\mathbf{x})$ is the diagonal matrix given by

$$\mathbf{D}^{(L)}(\mathbf{x}) = \text{diag}\left(\dot{\sigma}\left(f_i^{(L)}(\mathbf{x})\right)\right). \tag{15.72}$$

We introduce the notation $b^{(L)}(\mathbf{x}) := \sqrt{c_\sigma/d_L} \mathbf{D}^{(L)}(\mathbf{x}) \left(\mathbf{W}^{(L+1)}\right)^\top$, which is used later.

Next, note that

$$\frac{\partial f(\mathbf{x}; \boldsymbol{\theta})}{\partial \mathbf{W}^{(L-1)}} = \sum_{i=1}^{d_L} \underbrace{\frac{\partial f(\mathbf{x}; \boldsymbol{\theta})}{\partial g_i^{(L)}(\mathbf{x})} \frac{\partial g_i^{(L)}(\mathbf{x})}{\partial f_i^{(L)}(\mathbf{x})}}_{b_i^{(L)}(\mathbf{x})} \frac{\partial f_i^{(L)}(\mathbf{x})}{\partial \mathbf{W}^{(L-1)}}. \tag{15.73}$$

Here, $\dfrac{\partial f(\mathbf{x};\boldsymbol{\theta})}{\partial g_i^{(L)}(\mathbf{x})}\dfrac{\partial g_i^{(L)}(\mathbf{x})}{\partial f_i^{(L)}(\mathbf{x})}$ is exactly the i-th entry of $b^{(L)}(\mathbf{x})$. Moreover, the relationship between $f_i^{(L)}(\mathbf{x})$ and $\mathbf{W}^{(L-1)}$ is basically the same as the relationship between $f(\mathbf{x};\boldsymbol{\theta}) = f^{(L+1)}(\mathbf{x})$ and $\mathbf{W}^{(L)}$. Therefore, it is easy to derive

$$\frac{\partial f_i^{(L)}(\mathbf{x})}{\partial \mathbf{W}^{(L-1)}} = \sqrt{\frac{c_\sigma}{d_{L-1}}}\mathbf{D}^{(L-1)}(\mathbf{x})\left(\mathbf{W}_i^{(L)}\right)^\top \left(g^{(L-2)}(\mathbf{x})\right)^\top, \quad (15.74)$$

where

$$\mathbf{D}^{(L-1)}(\mathbf{x}) = \mathrm{diag}\left(\dot{\sigma}\left(f_i^{(L-1)}(\mathbf{x})\right)\right). \quad (15.75)$$

Combining (15.73) and (15.74), we have

$$\frac{\partial f(\mathbf{x};\boldsymbol{\theta})}{\partial \mathbf{W}^{(L-1)}} = \sum_{i=1}^{d_{L-1}} b_i^{(L)}(\mathbf{x})\sqrt{\frac{c_\sigma}{d_{L-1}}}\mathbf{D}^{(L-1)}(\mathbf{x})\left(\mathbf{W}_i^{(L)}\right)^\top \left(g^{(L-2)}(\mathbf{x})\right)^\top$$

$$(15.76)$$

$$= \sqrt{\frac{c_\sigma}{d_{L-1}}}\underbrace{\mathbf{D}^{(L-1)}(\mathbf{x})}_{\in\mathbb{R}^{d_{L-1}\times d_{L-1}}}\overbrace{\underbrace{\left(\mathbf{W}^{(L)}\right)^\top}_{\in\mathbb{R}^{d_{L-1}\times d_L}}\underbrace{b^{(L)}(\mathbf{x})}_{\in\mathbb{R}^{d_L\times 1}}}^{=:b^{(L-1)}(\mathbf{x})}\underbrace{\left(g^{(L-2)}(\mathbf{x})\right)^\top}_{\in\mathbb{R}^{1\times d_L}}.$$

$$(15.77)$$

We see that (15.77) has the same form of (15.71). Let's define recursively $b^{(L+1)}(\mathbf{x}) \equiv 1$ and

$$b^{(h)}(\mathbf{x}) = \sqrt{\frac{c_\sigma}{d_h}}\mathbf{D}^{(h)}(\mathbf{x})(\mathbf{W}^{(h+1)})^\top b^{(h+1)}(\mathbf{x}), \quad (15.78)$$

for $h = L, L-1, \ldots, 2, 1$, where

$$\mathbf{D}^{(h)}(\mathbf{x}) = \mathrm{diag}\left(\dot{\sigma}\left(f_i^{(h)}(\mathbf{x})\right)\right). \quad (15.79)$$

By induction,

$$\frac{\partial f(\mathbf{x};\boldsymbol{\theta})}{\partial \mathbf{W}^{(h)}} = b^{(h)}(\mathbf{x})\left(g^{(h-1)}(\mathbf{x})\right)^\top, \quad h = 1, 2, \ldots, L, L+1. \quad (15.80)$$

Now that we have an expression of $\frac{\partial f(\mathbf{x};\boldsymbol{\theta})}{\partial \mathbf{W}^{(h)}}$ for all h, we are ready to calculate $\left\langle \frac{\partial f(\mathbf{x};\boldsymbol{\theta})}{\partial \boldsymbol{\theta}}, \frac{\partial f(\mathbf{x}';\boldsymbol{\theta})}{\partial \boldsymbol{\theta}} \right\rangle = \sum_{h=1}^{L+1} \left\langle \frac{\partial f(\mathbf{x};\boldsymbol{\theta})}{\partial \mathbf{W}^{(h)}}, \frac{\partial f(\mathbf{x}';\boldsymbol{\theta})}{\partial \mathbf{W}^{(h)}} \right\rangle$. Specifically, we write

$$\left\langle \frac{\partial f(\mathbf{x};\boldsymbol{\theta})}{\partial \mathbf{W}^{(h)}}, \frac{\partial f(\mathbf{x}';\boldsymbol{\theta})}{\partial \mathbf{W}^{(h)}} \right\rangle$$

$$= \left\langle b^{(h)}(\mathbf{x}) \left(g^{(h-1)}(\mathbf{x}) \right)^{\top}, b^{(h)}(\mathbf{x}') \left(g^{(h-1)}(\mathbf{x}') \right)^{\top} \right\rangle \tag{15.81}$$

$$= \mathrm{tr}\left(g^{(h-1)}(\mathbf{x}) \left(b^{(h)}(\mathbf{x}) \right)^{\top} b^{(h)}(\mathbf{x}') \left(g^{(h-1)}(\mathbf{x}') \right)^{\top} \right) \tag{15.82}$$

$$= \mathrm{tr}\left(\left(b^{(h)}(\mathbf{x}) \right)^{\top} b^{(h)}(\mathbf{x}') \left(g^{(h-1)}(\mathbf{x}') \right)^{\top} g^{(h-1)}(\mathbf{x}) \right) \tag{15.83}$$

$$= \left(b^{(h)}(\mathbf{x}) \right)^{\top} b^{(h)}(\mathbf{x}') \left(g^{(h-1)}(\mathbf{x}') \right)^{\top} g^{(h-1)}(\mathbf{x}) \tag{15.84}$$

$$= \left\langle b^{(h)}(\mathbf{x}), b^{(h)}(\mathbf{x}') \right\rangle \left\langle g^{(h-1)}(\mathbf{x}), g^{(h-1)}(\mathbf{x}') \right\rangle, \tag{15.85}$$

where the inner product in (15.81) is the Hilbert–Schmidt inner product. For two real matrices $\mathbf{A}, \mathbf{B}$ of the same shape, it is defined by

$$\langle \mathbf{A}, \mathbf{B} \rangle := \sum_{i,j} A_{ij} B_{ij} = \mathrm{tr}\left(\mathbf{A}^{\top} \mathbf{B} \right). \tag{15.86}$$

On one hand, $\left\langle g^{(h-1)}(\mathbf{x}), g^{(h-1)}(\mathbf{x}') \right\rangle$ is described by NN-GP from Chapter 15. On the other hand, for $h = L, L-1, \ldots, 2, 1$,

$$\left\langle b^{(h)}(\mathbf{x}), b^{(h)}(\mathbf{x}') \right\rangle$$

$$= \left\langle \underbrace{\sqrt{\frac{c_\sigma}{d_h}} \mathbf{D}^{(h)}(\mathbf{x}) \left(\mathbf{W}^{(h+1)} \right)^{\top}}_{=: \mathbf{W}^{\top}} \underbrace{b^{(h+1)}(\mathbf{x})}_{=:b}, \right.$$

$$\left. \underbrace{\sqrt{\frac{c_\sigma}{d_h}} \mathbf{D}^{(h)}(\mathbf{x}') \left(\mathbf{W}^{(h+1)} \right)^{\top}}_{=: \mathbf{W}^{\top}} \underbrace{b^{(h+1)}(\mathbf{x}')}_{=:b'} \right\rangle \tag{15.87}$$

$$= \frac{c_\sigma}{d_h} \sum_i \left(\dot{\sigma} \left(f_i^{(h)}(\mathbf{x}) \right) \sum_j W_{ji} b_j \right) \left(\dot{\sigma} \left(f_i^{(h)}(\mathbf{x}') \right) \sum_k W_{ki}' b_k' \right)$$

(15.88)

$$= \frac{c_\sigma}{d_h} \sum_i \dot{\sigma} \left(f_i^{(h)}(\mathbf{x}) \right) \dot{\sigma} \left(f_i^{(h)}(\mathbf{x}') \right) \sum_j \sum_k W_{ji} W_{ki} b_j b_k'. \quad (15.89)$$

The "trick" for proceeding from (15.89) is to replace W_{ji} and W_{ki} here with their expectation so that the sum $\sum_j \sum_k$ becomes independent of i. At infinite width, we can treat $\mathbf{W}^{(h+1)}$ and $b^{(h+1)}(\mathbf{x})$ as independent. We are not presenting the technical details here but assume these are valid. We refer to the work of Arora *et al.* (2019b) for readers interested in rigorous proofs.

By taking expectation, $\sum_j \sum_k W_{ji} W_{ki} b_j b_k'$ becomes $\sum_j b_j b_j'$. We thus have

$$\left\langle b^{(h)}(\mathbf{x}), b^{(h)}(\mathbf{x}') \right\rangle$$

$$= \frac{c_\sigma}{d_h} \sum_i \dot{\sigma} \left(f_i^{(h)}(\mathbf{x}) \right) \dot{\sigma} \left(f_i^{(h)}(\mathbf{x}') \right) \sum_j b_j b_j' \quad (15.90)$$

$$= c_\sigma \frac{1}{d_h} \sum_i \dot{\sigma} \left(f_i^{(h)}(\mathbf{x}) \right) \dot{\sigma} \left(f_i^{(h)}(\mathbf{x}') \right) \left\langle b^{(h+1)}(\mathbf{x}), b^{(h+1)}(\mathbf{x}') \right\rangle,$$

(15.91)

which becomes

$$c_\sigma \mathbb{E}_{(\mathbf{u},\mathbf{v}) \sim \mathcal{N}(\mathbf{0}, \mathbf{\Lambda}^{(h)})} \left[\dot{\sigma}(\mathbf{u}) \dot{\sigma}(\mathbf{v}) \right] \left\langle b^{(h+1)}(\mathbf{x}), b^{(h+1)}(\mathbf{x}') \right\rangle \quad (15.92)$$

at infinite width. By defining

$$\dot{\mathbf{\Sigma}}^{(h)}(\mathbf{x}, \mathbf{x}') := c_\sigma \mathbb{E}_{(\mathbf{u},\mathbf{v}) \sim \mathcal{N}(\mathbf{0}, \mathbf{\Lambda}^{(h)})} \left[\dot{\sigma}(\mathbf{u}) \dot{\sigma}(\mathbf{v}) \right], \quad (15.93)$$

we conclude that

$$\left\langle b^{(h)}(\mathbf{x}), b^{(h)}(\mathbf{x}') \right\rangle = \prod_{h'=h}^{L} \dot{\mathbf{\Sigma}}^{(h)}(\mathbf{x}, \mathbf{x}'). \quad (15.94)$$

To summarize, we have

$$\mathbf{H}^*(\mathbf{x}, \mathbf{x}') = \left\langle \frac{\partial f(\mathbf{x}; \boldsymbol{\theta})}{\partial \boldsymbol{\theta}}, \frac{\partial f(\mathbf{x}'; \boldsymbol{\theta})}{\partial \boldsymbol{\theta}} \right\rangle \tag{15.95}$$

$$= \sum_{h=1}^{L+1} \left\langle \frac{\partial f(\mathbf{x}; \boldsymbol{\theta})}{\partial \mathbf{W}^{(h)}}, \frac{\partial f(\mathbf{x}'; \boldsymbol{\theta})}{\partial \mathbf{W}^{(h)}} \right\rangle \tag{15.96}$$

$$= \sum_{h=1}^{L+1} \left\langle b^{(h)}(\mathbf{x}), b^{(h)}(\mathbf{x}') \right\rangle \boldsymbol{\Sigma}^{(h-1)}(\mathbf{x}, \mathbf{x}') \tag{15.97}$$

$$= \boldsymbol{\Sigma}^{(L)}(\mathbf{x}, \mathbf{x}') + \sum_{h=1}^{L} \boldsymbol{\Sigma}^{(h-1)}(\mathbf{x}, \mathbf{x}') \prod_{h'=h}^{L} \dot{\boldsymbol{\Sigma}}^{(h')}(\mathbf{x}, \mathbf{x}'). \tag{15.98}$$

PART 6

Further Topics

Chapter 16

Transfer Learning

16.1 Overview

In real-world applications, we often need to deal with various machine learning tasks. Usually, models trained in one task do not work well for new tasks. In a standard machine learning scenario, we need to obtain large amounts of labeled data for each task. However, this is often difficult or expensive.

Transfer learning refers to machine learning techniques which extract knowledge from solving one problem or application and apply it to a different problem or application. Here, "different" refers to the fact that the datasets in the two problems are not the same, or do not follow the same distribution. The dataset used for training the model is called the *source domain*, while the dataset to which the model is applied to is called the *target domain*.

When we speak of a domain $\mathcal{D}$, we refer to a pair $\mathcal{D} = (\mathcal{X}, \mathbb{P})$, where $\mathcal{X}$ is the feature space that contains the data points (so that each data point $\mathbf{x} \in \mathcal{X}$) and $\mathbb{P}$ is the marginal probability distribution over the feature space. Therefore, when we say the source domain and the target domain are different, there are two cases. First, they have different feature spaces; second, they have the same feature space but different distributions.

Given $\mathcal{D} = (\mathcal{X}, \mathbb{P})$, a task $\mathcal{T}$ is also a pair $\mathcal{T} = (\mathcal{Y}, f)$, where $\mathcal{Y}$ is a label space (so that each label $y \in \mathcal{Y}$) and $f : \mathcal{X} \to \mathcal{Y}$ is a function that predicts the labels of unseen data. Under this terminology, given a source domain $\mathcal{D}_s$ and task $\mathcal{T}_s$, a target domain $\mathcal{D}_t$ and task $\mathcal{T}_t$, *transfer learning* refers to learning a good function f_t for the target domain by extracting information from $\mathcal{D}_s$ and $\mathcal{T}_s$, where $\mathcal{D}_s \neq \mathcal{D}_t$ or $\mathcal{T}_s \neq \mathcal{T}_t$.

There are four typical categories of transfer learning algorithms, namely instance-based, relation-based, model-based, and feature-based. Since we focus on deep learning, we only give a very quick review of instance- and relation-based methods but elaborate more on model- and feature-based methods.

In instance-based transfer learning, algorithms are mainly based on techniques including instance weighting, instance selection, or instance augmentation. Specifically, instance weighting refers to assigning weights to data points in the source domain based on their relevance to the target domain, and instance selection involves selecting a subset of data points in the source domain (setting zero weights to the remaining data). With instance weighting or instance selection, we hope that the model focuses on extracting the most important patterns for the target domain. Instance augmentation involves creating new instances based on the existing source domain instances so that the model generalizes better from the sources domain to the target domain. Examples of instance augmentation include rotating or flipping images, or adding noise to the data.

In relation-based transfer learning, the machine learning model is trained on a source task that has some known relations or structures, and then the learned knowledge is transferred to a target task that also has similar relations or structures. The goal is to leverage the similarities between the two tasks to improve the performance of the model on the target task. There are several techniques that can be used for relation-based transfer learning, such as sharing weights between the source and target models, adapting the model architecture to the target task, and using transferable features that are common to both tasks. The choice of technique depends on the specific application and the similarity between the source and target tasks.

16.2 Model-Based Transfer Learning

In model-based transfer learning, a model pretrained on a source domain is used for the task in the target domain. For deep learning models, the target domain does not have to be similar to the source domain. Indeed, many CNN and transformer models have pretrained weights and can be effectively used for a wide range of tasks. The pretrained model has already learned useful features or representations from a large amount of data in a source domain, and these features or representations can be fine-tuned to the target domain with a smaller amount of labeled data.

For instance, when using CNN for computer vision tasks, the pretrained CNN model is typically trained on a large dataset, such as ImageNet, which contains millions of images and thousands of classes. The pretrained model has already learned to extract useful features or representations from the images, such as edges and textures, and these features can be fine-tuned to a new task with a smaller amount of labeled data. As an example, we can keep the first several convolutional layers as feature extractors and train new layers for the downstream task. In other words, the convolutional layers we keep are believed to be general and applicable to a wide range of tasks; the deep layers are more task specific. When we train new layers, the convolutional layers we transfer from the pretrained model are frozen (weights fixed) or fine-tuned with a smaller learning rate.

The following is a PyTorch example of obtaining a pretrained ResNet model and changing the last fully connected (fc) layer. In this case, the pretrained model is trained on an image classification task. The target domain also contains a classification task but with a different number of classes. The last layer is modified so that the dimension of the output matches the number of classes in the target domain.

```python
import torch
import torch.nn as nn
from torchvision import models

model = models.resnet18(pretrained=use_pretrained)

for param in model.parameters():
    param.requires_grad = False

num_ft = model.fc.in_features
model.fc = nn.Linear(num_ft, num_classes)
```

Yosinski *et al.* (2014) studied the transferability of features in neural networks. Specifically, they created two classification tasks $\mathcal{A}$ and $\mathcal{B}$ by taking disjointed subsets of the ImageNet dataset and trained an eight-layer CNN, called base $\mathcal{A}$, on Task $\mathcal{A}$, and another eight-layer CNN, called base$\mathcal{B}$, on Task $\mathcal{B}$. Then, they chose a layer $n \in \{1, \ldots, 7\}$ and trained the following neural networks:

- $\mathcal{B}n\mathcal{B}$: Using the first n layers directly copied from base $\mathcal{B}$ and train the remaining layers on $\mathcal{B}$ with random initialized weights.
- $\mathcal{A}n\mathcal{B}$: Using the first n layers directly copied from base $\mathcal{A}$ and train the remaining layers on $\mathcal{B}$ with random initialized weights.
- $\mathcal{B}n\mathcal{B}^+$: The same as $\mathcal{B}n\mathcal{B}$ except for that the first n layers are further fine-tuned.
- $\mathcal{A}n\mathcal{B}^+$: The same as $\mathcal{A}n\mathcal{B}$ except for that the first n layers are further fine-tuned.

The top-1 accuracy (the conventional notion of "accuracy" in the context of the ImageNet dataset) reported by Yosinski *et al.* (2014) is shown in Figure 16.1.

The findings of Yosinski *et al.* (2014) suggested that early layers of a neural network tend to capture more generic features that can be transferred successfully, while deeper layers tend to specialize in task-specific information. The experiments conducted with the

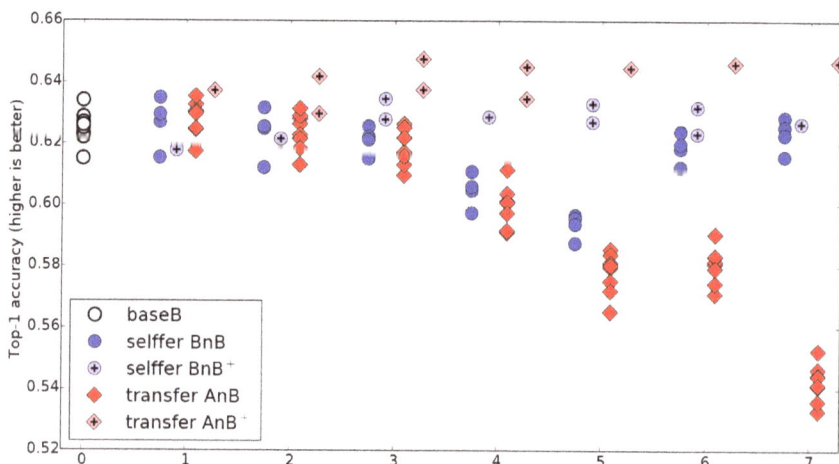

Fig. 16.1. The experimental results reported by Yosinski *et al.* (2014).

aforementioned network configurations provided empirical evidence supporting these observations.

16.3 Feature-Based Transfer Learning

In feature-based transfer learning, we train mapping functions ϕ_s and ϕ_t in the source and target domains, respectively. We map the data in both domains to a common feature space, which is said to be *domain invariant* since the distributions of features from source and target domains are identical or similar.

To quantify how similar two distributions are, we have already discussed using the KL divergence or the Wasserstein distance. In transfer learning literature, another measure is widely adopted: the *maximum mean discrepancy (MMD)*. In general, for two distributions P and Q, we define

$$\mathrm{MMD}(P, Q) = \|\mathbb{E}_{\mathbf{x} \sim P}(\phi(\mathbf{x})) - \mathbb{E}_{\mathbf{x} \sim Q}(\phi(\mathbf{x}))\|_{\mathcal{H}}, \tag{16.1}$$

where $\mathcal{H}$ is a real-valued reproducing kernel Hilbert space (RKHS). We only discuss these concepts in the context of transfer learning here while referring interested readers to the work of Gretton *et al.* (2012) and Hofmann *et al.* (2008) for comprehensive introductions.

Problem 16.1. In this exercise, we derive a dual form of the MMD.

(1) **Prove:** For any $f \in \mathcal{H}$,

$$\|f\|_{\mathcal{H}} = \sup_{\|g\|_{\mathcal{H}} \leq 1} \langle g, f \rangle_{\mathcal{H}}.$$

(2) **Prove:**

$$\mathrm{MMD}(P, Q) = \sup_{\|f\|_{\mathcal{H}} \leq 1} \mathbb{E}_{\mathbf{x} \sim P}(f(\mathbf{x})) - \mathbb{E}_{\mathbf{x} \sim Q}(f(\mathbf{x})).$$

Hint: Use Part (a) and the reproducing property: $\langle f, \phi(\mathbf{x}) \rangle_{\mathcal{H}} = f(\mathbf{x})$ for any $f \in \mathcal{H}$.

Note that the dual form looks very similar to the Wasserstein-1 distance used in WGAN, but the supremum is taken over a different set of f.

The Maximum Mean Discrepancy (MMD) between two domain samples $\mathbf{X}_s = \{\mathbf{x}_i^s\}_{i=1}^{n_s}, \mathbf{X}_t = \{\mathbf{x}_i^t\}_{i=1}^{n_t} \subset \mathbb{R}^D$ (regarded as empirical distributions) is defined to be

$$\text{MMD}(\mathbf{X}_s, \mathbf{X}_t) = \left\| \frac{1}{n_s} \sum_{i=1}^{n_s} \phi(\mathbf{x}_i^s) - \frac{1}{n_t} \sum_{i=1}^{n_t} \phi(\mathbf{x}_i^t) \right\|_{\mathcal{H}}. \tag{16.2}$$

A typical feature-based transfer learning algorithm will look for an identical feature extractor ϕ to map input from both source and target domains to the domain-invariant feature space and try to minimize the MMD between $\phi(\mathbf{X}_s)$ and $\phi(\mathbf{X}_t)$. That is, ϕ is the optimizer of

$$\min_{\phi} \text{MMD}(\phi(\mathbf{X}_s), \phi(\mathbf{X}_t)) + \lambda G(\phi), \tag{16.3}$$

where $\lambda G(\phi)$ is a regularization term.

Suppose the RKHS $\mathcal{H}$ is associated with the kernel k such that $k(\mathbf{x}, \mathbf{x}') = \langle \phi(\mathbf{x}), \phi(\mathbf{x}') \rangle_{\mathcal{H}}$. Define the $(n_s + n_t)$-by-$(n_s + n_t)$ matrix $\mathbf{K}$ to be

$$\mathbf{K} = \begin{bmatrix} \mathbf{K}_{s,s} & \mathbf{K}_{s,t} \\ \mathbf{K}_{s,t}^\top & \mathbf{K}_{t,t} \end{bmatrix}$$

in which the (i, j)th entry of $\mathbf{K}_{u,v}$ is given by $k(\mathbf{x}_i^u, \mathbf{x}_j^v)$ (where $u, v \in \{s, t\}$). Also, define a matrix $\mathbf{L}$ with the same shape as $\mathbf{K}$ to be

$$\mathbf{L} = \begin{bmatrix} \mathbf{L}_{s,s} & \mathbf{L}_{s,t} \\ \mathbf{L}_{s,t}^\top & \mathbf{L}_{t,t} \end{bmatrix},$$

where $\mathbf{L}_{s,s}$ is a block matrix (with the same shape as $\mathbf{K}_{s,s}$) whose entries are all $\frac{1}{n_s^2}$, $\mathbf{L}_{t,t}$ is a block matrix whose entries are all $\frac{1}{n_t^2}$, and $\mathbf{L}_{s,t}$ is a block matrix whose entries are all $-\frac{1}{n_s n_t}$. It is routine calculation to check that $\text{MMD}^2(\mathbf{X}_s, \mathbf{X}_t) = \text{tr}(\mathbf{KL})$. In view of this relation, we can represent (16.3) as

$$\min_{\phi} \text{tr}(\mathbf{KL}) + \lambda G(\phi). \tag{16.4}$$

16.4   Analysis of Transfer Learning

We assume that we have different source and target marginal distributions, but the same task. This setting is called *transductive* transfer learning, or *domain adaptation*. Specifically, we assume the same feature space $\mathcal{X}$ and the same label space $\mathcal{Y}$ for both the source and target domains. For simplicity, we consider the binary classification problem with a true labeling function $f : \mathcal{X} \to \{0, 1\}$. Let P denote the marginal distribution on the source domain and Q denote the marginal distribution on the target domain. The sample from the source domain $\{\mathbf{x}_i^s, y_i^s\}_{i=1}^{n_s}$ include pairs of instances and labels, while the sample from the target domain $\{\mathbf{x}_i^t\}_{i=1}^{n_t}$ may not include any label (unsupervised) or may include only few labeled data (semi-supervised).

In what follows, we show how the error in target domain compares with the training error in the source domain. Consider a hypothesis class $\mathcal{H}$ of candidates of labeling functions $h \in \mathcal{H}$. For any $h \in \mathcal{H}$, we define the true error in the source domain to be

$$\epsilon_s(h) = \mathbb{P}_P(f \neq h) \tag{16.5}$$

and the true error in the target domain to be

$$\epsilon_t(h) = \mathbb{P}_Q(f \neq h). \tag{16.6}$$

Correspondingly, we define the empirical error, or training error in the source and target domains, to be

$$\hat{\epsilon}_s(h) = \frac{1}{n_s} \sum_{i=1}^{n_s} \mathbf{1}(f(\mathbf{x}_i^s) \neq h(\mathbf{x}_i^s)) = \frac{1}{n_s} |\{i \in [n_s] : f(\mathbf{x}_i^s) \neq h(\mathbf{x}_i^s)\}| \tag{16.7}$$

and

$$\hat{\epsilon}_t(h) = \frac{1}{n_t} \sum_{i=1}^{n_t} \mathbf{1}(f(\mathbf{x}_i^t) \neq h(\mathbf{x}_i^t)) = \frac{1}{n_t} |\{i \in [n_t] : f(\mathbf{x}_i^t) \neq h(\mathbf{x}_i^t)\}|, \tag{16.8}$$

respectively. Moreover, we define the $\mathcal{H}$-divergence of P and Q, with respect to the hypothesis class $\mathcal{H}$, to be

$$d_{\mathcal{H}}(P, Q) = 2 \sup_{h \in \mathcal{H}} |\mathbb{P}_P \mathcal{I}(h) - \mathbb{P}_Q \mathcal{I}(h)|, \tag{16.9}$$

where $\mathcal{I}(h) = \{\mathbf{x} \in \mathcal{X} : h(\mathbf{x}) = 1\}$ is the set of instances with which h outputs 1 (the "positive class" determined by h). We have the following theorem.

Theorem 16.1. *Let $\mathcal{H}$ be a hypothesis class. Let*

$$h^* = \arg\min_{h \in \mathcal{H}} \epsilon_s(h) + \epsilon_t(h) \tag{16.10}$$

and

$$\lambda = \epsilon_s(h^*) + \epsilon_t(h^*). \tag{16.11}$$

Suppose $\epsilon_s(h) \le \hat{\epsilon}_s(h) + \mu$ for some μ. Then,

$$\epsilon_t(h) \le \hat{\epsilon}_s(h) + d_{\mathcal{H}}(P, Q) + \lambda + \mu. \tag{16.12}$$

Proof. We would like to upper bound

$$\epsilon_t(h) = \mathbb{P}_Q(f \ne h) = \mathbb{P}_Q(\{\mathbf{x} \in \mathcal{X} : f(\mathbf{x}) \ne h(\mathbf{x})\}). \tag{16.13}$$

Note that

$$\{\mathbf{x} \in \mathcal{X} : f(\mathbf{x}) \ne h(\mathbf{x})\} \tag{16.14}$$

$$\subset \{\mathbf{x} \in \mathcal{X} : h^*(\mathbf{x}) \ne f(\mathbf{x}) \text{ or } h^*(\mathbf{x}) \ne h(\mathbf{x})\} \tag{16.15}$$

$$= \{\mathbf{x} \in \mathcal{X} : h^*(\mathbf{x}) \ne f(\mathbf{x})\} \bigcup \{\mathbf{x} \in \mathcal{X} : h^*(\mathbf{x}) \ne h(\mathbf{x})\}. \tag{16.16}$$

Therefore,

$$\epsilon_t(h) = \mathbb{P}_Q(f \ne h) \tag{16.17}$$

$$\le \mathbb{P}_Q(h^* \ne f) + \mathbb{P}_Q(h^* \ne h) \tag{16.18}$$

$$\le \epsilon_t(h^*) + \mathbb{P}_P(h^* \ne h) + |\mathbb{P}_Q(h^* \ne h) - \mathbb{P}_P(h^* \ne h)| \tag{16.19}$$

$$\le \epsilon_t(h^*) + \mathbb{P}_P(h^* \ne h) + d_{\mathcal{H}}(P, Q) \tag{16.20}$$

$$\le \epsilon_t(h^*) + \mathbb{P}_P(h^* \ne f) + \mathbb{P}_P(h \ne f) + d_{\mathcal{H}}(P, Q) \tag{16.21}$$

$$= \epsilon_t(h^*) + \epsilon_s(h^*) + \epsilon_s(h) + d_{\mathcal{H}}(P, Q). \tag{16.22}$$

Since $\epsilon_s(h) \le \hat{\epsilon}_s(h) + \mu$, we have obtained (16.12). $\square$

We remark that $\epsilon_s(h) \leq \hat{\epsilon}_s(h) + \mu$ has to do with the generalization bound of training, which is studied in PAC learning theory and not discussed in this book. We refer interested readers to the work of Shalev-Shwartz and Ben-David (2014).

Let's discuss what implication can be drawn from the above result. In (16.12), the true error of h is split into four terms. In these four terms, λ is the smallest possible error and will not change with the algorithm. μ is the difference between the true and empirical errors in the source domain. In order to make μ small, we need good performance of generalization in the source domain. To control the remaining two terms, we need both good classification performance on the training data, as well as representations from the source and target domains that are as indistinguishable as possible. This agrees with our discussion in the previous chapter on feature-based transfer learning methods.

16.5   Domain Adversarial Neural Network

A foundational transfer learning algorithm following the above idea is the Domain Adversarial Neural Network (DANN) proposed by Ganin *et al.* (2016). Both training samples $\mathbf{x}^s$ from the source domain and training samples $\mathbf{x}^t$ from the target domain are processed by a feature extractor G to produce features $\mathbf{h}^s$ and $\mathbf{h}^t$, respectively. A label classifier C processes $\mathbf{h}^s$ and provides a classification loss ℓ_C by comparing its output $\hat{\mathbf{y}}^s$ with labels $\mathbf{y}^s$ from the source domain. At the same time, a domain discriminator D produces a loss ℓ_D by comparing $\mathbf{h}^s$ and $\mathbf{h}^t$. An illustration of DANN is given in Figure 16.2.

What we need to implement is a training algorithm that determines G, C, and D. It is clear how the two losses ℓ_C and ℓ_D depend on C and D, respectively. Nevertheless, it is important to note that the domain discriminator is used to promote a close $\mathbf{h}^t$ with $\mathbf{h}^s$, which are outputs of G. Therefore, the role G plays is to "fool" the domain discriminator. This is similar to how the generator tries to fool the discriminator in GAN (recall Chapter 11). Therefore, the optimization problem for G, C, and D is again a min-max problem. Given pairs of training examples and labels $\{\mathbf{x}_i^s, y_i^s\}_{i=1}^{n_s}$ from the source domain, and training examples $\{\mathbf{x}_i^t\}_{i=1}^{n_t}$ from the target domain, the overall training objective is

$$\min_{\boldsymbol{\theta}_G, \boldsymbol{\theta}_C} \max_{\boldsymbol{\theta}_D} L(\boldsymbol{\theta}_G, \boldsymbol{\theta}_C, \boldsymbol{\theta}_D), \tag{16.23}$$

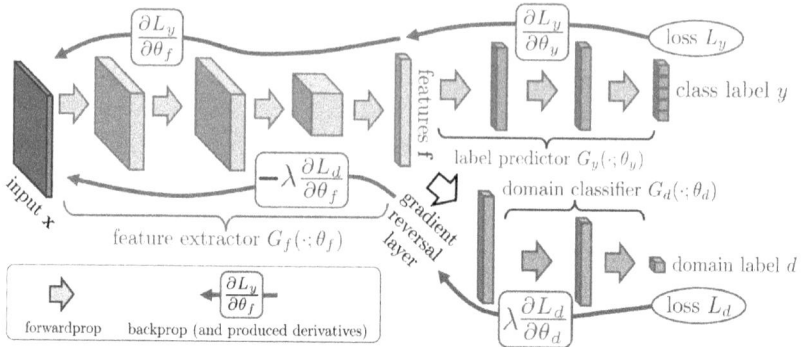

Fig. 16.2. An illustration of DANN. The image is taken from the work of Ganin *et al.* (2016).

where $\boldsymbol{\theta}_G$, $\boldsymbol{\theta}_C$, $\boldsymbol{\theta}_D$ are the parameters of G, C, D, respectively, and

$$L(G, C, D) = \frac{1}{n_s} \sum_{i=1}^{n_s} \ell_C(\mathbf{x}_i^s, y_i^s; \boldsymbol{\theta}_G, \boldsymbol{\theta}_C)$$

$$- \lambda \left(\frac{1}{n_s} \ell_D(\mathbf{x}_i^s, y_i^s; \boldsymbol{\theta}_G, \boldsymbol{\theta}_D) + \frac{1}{n_t} \ell_D(\mathbf{x}_i^t, y_i^t; \boldsymbol{\theta}_G, \boldsymbol{\theta}_D) \right).$$

$$(16.24)$$

In DANN, there is a gradient reversal step for the different roles of G in ℓ_C and ℓ_D. We explain this as follows. Since $\boldsymbol{\theta}_G$ needs to minimize ℓ_C while maximize ℓ_D, it receives the gradients from C as usual in backpropagation, but reverses the gradients received from D. A PyTorch implementation of this layer can be found in, e.g., https://torcheeg.readthedocs.io/en/latest/torcheeg.trainers. html#trainers-danntrainer, where the gradient reversal layer is customized as follows:

```
from torch.autograd import Function

class GradientReverse(Function):
    @staticmethod
    def forward(ctx, x, lambda):
        ctx.lambda = lambda
        return x.view_as(x)

    @staticmethod
    def backward(ctx, grad_output):
        output = grad_output.neg() * ctx.lambda
        return output, None
```

Problem 16.2. The total variation (TV) distance between two probability distributions $\mathcal{D}$, $\mathcal{D}'$ is defined by

$$\mathrm{TV}(\mathcal{D}, \mathcal{D}') = 2 \sup_{E \in \mathcal{B}} |\mathbb{P}_\mathcal{D}(E) - \mathbb{P}_{\mathcal{D}'}(E)|,$$

where $\mathcal{B}$ is a set of events measurable by $\mathcal{D}$ and $\mathcal{D}'$, $\mathbb{P}_\mathcal{D}(E)$ is the probability of the event E under the distribution $\mathcal{D}$, and similarly is $\mathbb{P}_{\mathcal{D}'}(E)$ defined. (Remark: In literature, you might find the TV distance defined without the factor "2".)

(1) Prove: If $\mathcal{B}$ is the collection of all subsets of a finite set Ω, then

$$\mathrm{TV}(\mathcal{D}, \mathcal{D}') = \sum_{x \in \Omega} |\mathbb{P}_\mathcal{D}(x) - \mathbb{P}_{\mathcal{D}'}(x)|.$$

(2) Recall that in the theory of transfer learning, the $\mathcal{H}$-divergence is defined by

$$d_\mathcal{H}(\mathcal{D}, \mathcal{D}') = 2 \sup_{h \in \mathcal{H}} |\mathbb{P}_\mathcal{D}(I(h)) - \mathbb{P}_{\mathcal{D}'}(I(h))|,$$

where $x \in I(h) \Leftrightarrow h(x) = 1$. Can $d_\mathcal{H}$ be written in the form of a TV distance? If yes, specify $\mathcal{B}$.

(3) Assume $\mathbb{P}_\mathcal{D}$ and $\mathbb{P}_{\mathcal{D}'}$ adopt densities $p_\mathcal{D}$ and $p_{\mathcal{D}'}$, respectively. Prove that the TV distance is a Wasserstein distance in the sense that

$$\frac{1}{2} \mathrm{TV}(\mathcal{D}, \mathcal{D}') = \min_{\pi \in \Pi(p_\mathcal{D}, p_{\mathcal{D}'})} \mathbb{E}_\pi d(x, y),$$

where $d(x, y) = \mathbf{1}_{x \neq y} = \begin{cases} 1 & \text{if } x \neq y \\ 0 & \text{if } x = y \end{cases}$.

Chapter 17

Explainable AI

One of the strengths of neural networks, which also causes a difficulty in adopting them for critical applications, is the black-box nature of its internal workings. Built as a complex model with very many parameters that are tuned by exposing it to huge amounts of data, it is difficult for the human users to understand or justify the decisions made by such systems. The task of the machine learning researcher is in designing the architecture, defining the optimization criteria, and preparing the data rather than modeling or analyzing the problem itself. Unlike traditional modeling approach that starts with building a parametric model that approximates the mechanisms pertinent to the engineering problems at hand, the common practice in neural networks is to bypasses the modeling step by allowing huge amount of data to tune the many model parameters until some optimality criteria is satisfied. This approach results in systems whose operation may either defy the more traditional engineering intuition or simply do not require it. When such networks are tasked with high stake problems such as medical scoring, or socially sensitive issue such as credit scoring, among others, it becomes important to be able to explain the model decisions to the operators and end-users, who might be eventually affected by the system results. In many cases, the question of importance is less figuring out the actual internal operation of the network but rather finding out what aspects of the input data might have caused the machine intelligence to reach its decisions. In other words, one might feel more confident in accepting machine learning predictions if the results can be attributed to specific features of the data, done in ways that can be approximated by some logical considerations, such as finding equivalent derivations in terms of a linear combination of features according to an importance ranking. In this chapter, we cover the so-called "additive feature attribution method" that approximates the original complex prediction model by another, more simple linear explanation model. According to this approach, the best explanation of a simple model is the model itself since it perfectly represents itself and is easy to understand. Accordingly, we define an explanation model as any interpretable approximation of the original model. Beyond providing explanations to end-users, such explanation models might be useful for debugging the complex model predictions, analyzing its robustness, or extracting simple rules from the complex model. The so-called Shapley value-based explanation techniques (Lundberg and Lee, 2017) is a unified approach that use game-theoretical setting that considers Machine Learning task in terms of a coalition game as follows: The input features x are viewed as the players, the game payoff is viewed as the complex model outcome $f(x)$, and the explanation function

$$g(z') = \phi_0 + \sum_{j=1}^{M} \phi_j z'_j \qquad (17.1)$$

is a simplified linear function over a set of binary variable $z' \in \{0,1\}^M$ that represent the different coalitions with an entry of 1 meaning that the corresponding feature value is "present" and 0 that it is "absent", M is the number of features or maximum coalition size, and $\phi_j \in \mathbb{R}$ is the feature attribution for a feature j. Local explanation methods provide feature attributes per input instance x, so as to "explain" the specific prediction $f(x)$. In such a case, all features are present, which is often denoted as specific type of z', $x' = 1^M$, so that the formula simplifies to

$$g(x') = \phi_0 + \sum_{j=1}^{M} \phi_j. \qquad (17.2)$$

To get from coalitions of feature values to valid data instances, we need a function $x = h_x(x')$ that maps 1's from the coalition vector to the corresponding value from the instance x. Local methods try to ensure that $g(z') \approx f(h_x(z'))$ for cases where some of the input attributes are missing. Since the model was not trained on such specific examples, the core task of the explanation methods is to "simulate" the cases where some of the inputs are missing or unknown. The "explanations", also known as feature importance, are the values ϕ_j that will be computed as weighted average of all possible outcome of function $f(z')$ for coalitions or features sets that contain or are absent feature j. In the following section, we explain how the classic Shapley values are used to assign the weights for different feature coalitions. Next, we discuss some sampling and kernel regression methods to solve the explanation model estimation problem.

17.1 Shapley Value

Shapley value is a method from coalitional game theory that tries to solve the problem of how to fairly distribute the "payout" among the players once a certain game outcome is reached. To determine Shapley value, we consider a cooperative game played by a set of players $1, \ldots, M$. The game is defined by set function $v : 2^M \to R$ such that $v(S)$ is the payoff for any coalition of players S in $1, \ldots, M$, and $v = 0$. The task is to find a fair attributions (give a score) to each player that can be higher or lower than individual payoff depending on the game. Shapley value attributions are built by averaging the marginal contribution of a player to each possible coalition S.

Conceptually, the Shapley value solves the following different cases:

- The values of the coalitions are equal or greater to sum of individual values (fair allocation that takes into account synergy).
- The sum of the values of non-negative agents is greater than the total payoff of the grand coalition.
- Use of Shapley value in SHAP allows negative values (some agents take away from the total value).

An interesting application of the Shapley allocation is distribution of assets in a situation where the demands of the individual agents is greater then the total assets of the grand coalition, or in other words, the total worth of the assets (property) to be distribution, such as in the case of a bankruptcy problem. To give a concrete examples, let's consider two classical problems of asset allocation: one known

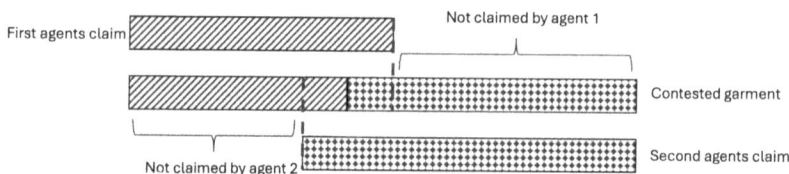

Fig. 17.1. The contested garment principle.

as the "Contested Garment Problem" from the Jewish Talmud and another is "Run to the Bank".

Talmud (Baba Mezi'a 2a, Chapter I, Babylonian Talmud) presented the following dilemma: Two [persons appearing before a court] hold a garment. If one says, "it is all mine", and the other says, "half of it is mine", the former then receives three quarters [of the value of the garment] and the latter receives one quarter. The explanation for this proposed solution is shown in Figure 17.1.

The principle of allocating the values to each contestant is as follows: The empty coalition has no value, then the first contestant claims that their value is the whole garment ("its all mine") while the second contestant claims only half. The total value of the assets do be dividing is a single garment, so the value for the joining coalition of first and second claimant is the whole garment as well. Before proceeding with the calculation of Shapley value that is explained in the following section, we try to develop an intuitive logic for a just division. The first claimant claims the whole garment, while the second claimant claims only half. Under this setting, the second claimant gave up half of the garment to the first claimant as they had no claim on it. Thus, the only contested part is half of the garment, for which there is equal claim of the first and second claimant. Accordingly, a fair division of the contested part is equal division; this gives one quarter of the garment to each claimant. This results in a division of three quarters versus one quarter to the first and second claimant, respectively. One may also want to compare the solution to Shapley value calculator that is available online at http://shapleyvalue.com/index.php.

A further intuition about Shapley value can be developed using the so-called "run to the bank" scenario (Young, 1995). The claimants arrive randomly in the race to the bank to withdraw their claims. The first to arrive takes his full claim, which may be equal to the entire estate if what is available is equal to or less than his claim.

Whoever arrives second does the same, and he may receive zero if nothing is left, and so on. Assuming the order of arrival is random, all orders are equally probable so that each claimant finally receives the expectation of all cases. According to this logic, in the case of the order first claimant, second claimant, the division is 1,0, while in the second order of arrival second claimant, first claimant, the division is 0.5, 0.5. Averaging the two gives final allocation of 0.75, 0.25.

17.2 Computing Shapley Values

There are two approaches to computing Shapley Values for a cooperative game: subsets-based and permutation-based. The game is played by a set of players $\mathcal{M} = \{1 \ldots M\}$ and is characterized by a set function $v : 2^M \rightarrow R$, such that $v(S)$ is the payoff for a coalition of players $S \subseteq \mathcal{M}$, and $v(\emptyset) = 0$. The subsets-based Shapley values are built by examining the marginal contribution of a player i, $v(S \cup \{i\}) - v(S)$, for an existing coalition S. The Shapley value of a player i is

$$\phi_i(v).$$

Permutation-based approach is using sampling from an ordered set of permutations πM. For a sample O, the order of the players (recall the "run to the bank" interpretation) can be examined by comparing the set of players that precede player i, $pre_i(O)$, and the value derived when player i is added. The complexity of the two approaches is $\mathbf{O}(2^M)$ and $\mathbf{O}(M!)$, respectively:

- Subsets-based approach, with complexity $\mathbf{O}(2^M)$

$$\phi_i(v) = \frac{1}{M} \sum_{S \subseteq \mathcal{M} \setminus \{i\}} \binom{M-1}{|S|}^{-1} (v(S \cup \{i\}) - v(S)). \quad (17.3)$$

- Permutation-based approach with complexity $\mathbf{O}(M!)$

$$\phi_i(v) = \mathop{\mathbb{E}}_{S \sim \pi(M)} [v(\mathrm{pre}_i(S) \cup \{i\}) - v(\mathrm{pre}_i(S))]. \quad (17.4)$$

It helps demonstrate the two approaches by a simple example. Lets assume a game characterized by the following function:

$v(\{1\}) = 100, v(\{2\}) = 125, v(\{3\}) = 50, v(\{1,2\}) = 270, v(\{1,3\}) = 375, v(\{2,3\}) = 350, v(\{1,2,3\}) = 500$. The following table, shown in

Probability	Order of arrival	1's marginal contribution	2's marginal contribution	3's marginal contribution
$\frac{1}{6}$	first 1 then 2 then 3: 123	v({1}) = 100	v({1,2}) − v({1}) = 270 − 100 = 170	v({1,2,3}) − v({1,2}) = 500 − 270 = 230
$\frac{1}{6}$	first 1 then 3 then 2: 132	v({1}) = 100	v({1,2,3}) − v({1,3}) = 500 − 375 = 125	v({1,3}) − v({1}) = 375 − 100 = 275
$\frac{1}{6}$	first 2 then 1 then 3: 213	v({1,2}) − v({2}) = 270 − 125 = 145	v({2}) =125	v({1,2,3}) − v({1,2}) = 500 − 270 = 230
$\frac{1}{6}$	first 2 then 3 then 1: 231	v({1,2,3}) − v({2,3}) = 500 − 350 = 150	v({2}) =125	v({2,3}) − v({2}) = 350 − 125 = 225
$\frac{1}{6}$	first 3 then 1 then 2: 312	v({1,3}) − v({3}) = 375 − 50 = 325	v({1,2,3}) − v({1,3}) = 500 − 375 = 125	v({3}) = 50
$\frac{1}{6}$	first 3 then 2 then 1: 321	v({1,2,3}) − v({2,3}) = 500 − 350 = 150	v({2,3}) − v({3}) = 350 − 50 = 300	v({3}) = 50

Fig. 17.2. The contested garment principle.

Figure 17.2 summarizes the marginal values as derive by considering the order of arrival and the remaining marginal contribution in each order.

Solving the problem using the subset approach gives the following solutions: For $i = 1$, the sets minus i are $S \setminus i = \{\emptyset\}, \{2\}, \{3\}, \{2, 3\}$. The subset weights $\binom{M-1}{|S|}$ for $M = 3$ are $\binom{2}{1} = 2$, $\binom{2}{2} = 1$, $\binom{2}{0} = 1$.

Summing the weights according to (17.3) results in

$$\phi_1 = 1/3(100 + 1/2 \cdot 145 + 1/2 \cdot 325 + 150) = 970/6. \qquad (17.5)$$

Using permutation solution results in

$$\phi_1 = 1/6(100 + 100 + 145 + 325 + 150). \qquad (17.6)$$

17.3 Additive Feature Attribution

To relate Shapley value to additive feature attribution, we require the following three properties:

(1) **Local accuracy:** $f(x) = \phi_0 + \sum_{j=1}^{M} \phi_j x_j' = \mathbb{E}_X(f(X)) + \sum_{j=1}^{M} \phi_j$, where $x' = (1, 1, \ldots 1)$.
(2) **Missingness:** $x_j' = 0 \Rightarrow \phi_j = 0$.

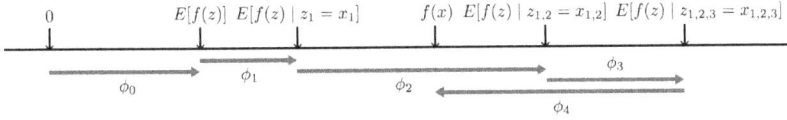

Fig. 17.3. SHAP values that explain how various feature contribute positively or negatively to get from the base value $\mathbb{E}[f(z)]$ to the current output value $f(x)$.

(3) **Consistency:** For any two models f and f', if

$$f'(z') - f'(z' \setminus i) \geq f(z') - f(z' \setminus i) \qquad (17.7)$$

for all inputs $z' \in \{0,1\}^M$, then $\phi_i(f') \geq \phi_i(f)$.

It can be shown that only one possible additive explanation model satisfies these properties with Shapley values as the weighting coefficient (Lundberg and Lee, 2017):

$$\phi_i(f) = \sum_{z' \cup x'} \frac{|z'|!(M - |z'| - i)!}{M!}[f(z') - f(z' \setminus i)], \qquad (17.8)$$

where $|z'|$ is the number of non-zero entries in z' and $z' \cup x'$ represents all binary z' vectors where the non-zero entries are a subset of the non-zero entries in x'. This result implies that models that do not satisfy the Shapley values violate the local accuracy and/or consistency.

A solution to (17.8) proposed in the work of Lundberg and Lee (2017) is defining the partial mappings $f(z')$ as conditional expectations on z_S, where S is the set on non-zero indexes in z'. Since most models cannot handle arbitrary patterns of missing input values, the proposed solution approximates $f(z_S)$ with $\mathbb{E}[f(z)|z_S]$. The resulting Shapley Additive Explanation (SHAP) attributes to each feature the change in the expected model prediction when conditioning on that feature. This process can be demonstrated for a particular ordering in Figure 17.3.

17.4　Kernel SHAP

Instead of solving the Shapley values, Kernel SHAP method solves the following optimization problem: Find g that minimizes the loss

$$L(f, g, \phi_{x'}) = \sum_{z' \in Z} [f(h_x(z')) - g(z')]^2 \pi_{x'}(z'), \qquad (17.9)$$

with a weight kernel

$$\pi_{x'}(z') = \frac{(M-1)}{\binom{M}{|z'|}|z'|(M-|z'|)}. \qquad (17.10)$$

Sketch of a proof that this kernel converges to Shapley values is as follows:

Let X be the matrix of all possible binary vectors with 2^M rows and M columns. Use Shapley kernel to solve the weighted linear regression for $y_i = f_x(S_i)$:

$$\phi = (X^T W X)^{-1} X^T W y. \qquad (17.11)$$

W is a diagonal matrix with the Shapley kernel weights for each row of X:

$$k(z') = k(M, s) = \frac{(M-1)}{\binom{M}{s})s(M-s)}. \qquad (17.12)$$

Using the approximation $(XWX)^{-1} = I + \frac{1}{M-1}(I - J)$, where I is the identity matrix and J is the matrix of all ones, one can show (see supplement materials, Lundberg and Lee, 2017) that

$$\phi_j = \sum_{S \subseteq N \setminus j} \frac{(M - s_i - 1)! s_i!}{M!} f_x(S \cup \{i\}) - f_x(S)]. \qquad (17.13)$$

Chapter 18

Deep Reinforcement Learning

18.1 Reinforcement Learning

Many real-world problems, such as game playing, robotics, and autonomous driving, involve decision making under complex and dynamic environments characterized by uncertainty, incompleteness, and high dimensionality. For instance, in game playing, the environment involves the state of the game, the move of the opponent, and the potential outcomes. Supervised or unsupervised learning methods do not fit to solve these kinds of problems. Instead, reinforcement learning (RL) involves an agent learning to make a sequence of decisions in an environment. The agent interacts with the environment through a series of actions and receives feedback in the form of rewards or punishments, depending on the actions taken. Nowadays, RL has developed to be a very important subfield of machine learning. For a comprehensive introduction to RL, we recommend two excellent resources: the textbook by Sutton and Barto (2018) and the lectures given by David Silver (2015).

A typical reinforcement learning algorithm involves an agent that interacts with an environment in a series of discrete time steps. At each time step, the agent receives an observation of the current state of the environment and selects an action to perform based on its current policy. The environment then transitions to a new state, and the agent receives a scalar reward signal indicating how well it performed in the previous state.

A typical RL scenario is illustrated in Figure 18.1. In this figure, at each time step t, the agent receives the state S_t and the reward

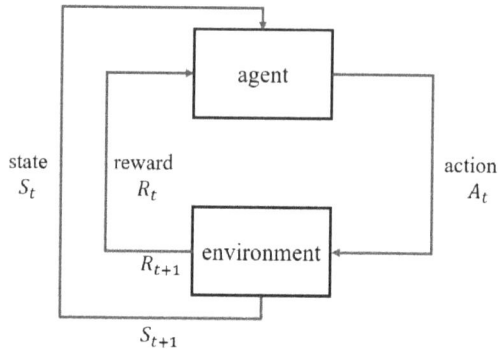

Fig. 18.1. A typical RL scenario.

R_t from the environment, and then takes an action A_t accordingly, which changes the environment, which produces a state S_{t+1} and R_{t+1} for the next time step.

18.2 Markov Decision Processes

The Markov Decision Process (MDP) is a mathematical framework used to model sequential decision-making problems, where the outcomes of an agent's actions depend on the current state of the environment, and the agent's objective is to maximize a cumulative reward over time. In an MDP, the agent's actions cause the environment to transition to a new state, and the agent receives a reward or punishment based on the new state. The agent's goal is to learn a policy, which is a mapping from states to actions, that maximizes the cumulative reward over time.

Let's first look at a very simple Markov chain that describes various states of pet cats: hunting, hungry, affectionate, confused, relaxed, and sleeping (well, cats may show much more different states but let's just use this many for our example). The Markov chain is illustrated in Figure 18.2. Each state in the chain represents a possible behavior of the cat, and the transition probabilities between them reflect how likely it is for a cat to switch from one behavior to another. For instance, if a cat is relaxed, it may become confused if its owner talks to it in human languages. The human talking is a possible action involved in reinforcement learning.

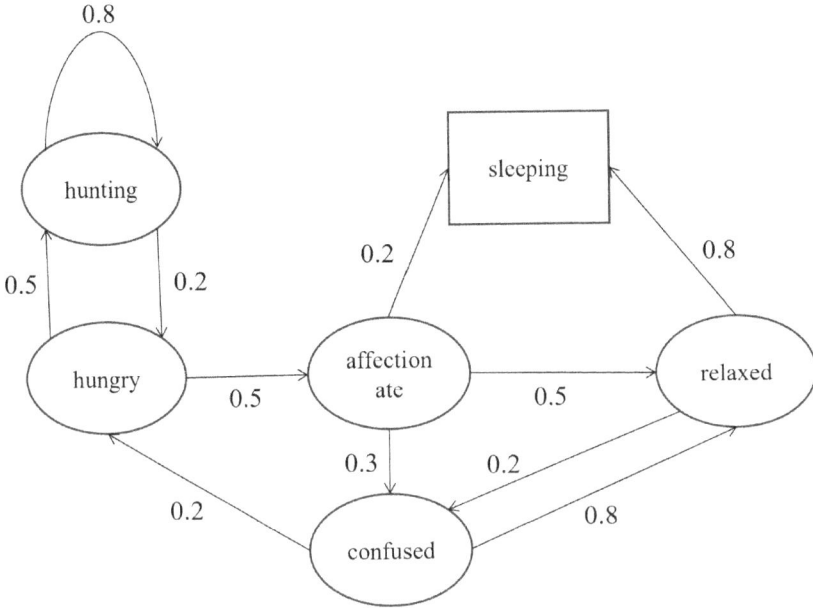

Fig. 18.2. A Markov chain that describes states of cats and the transition probabilities.

For the cat owner, each state of the cat corresponds to different levels of reward. For example, the state of being affectionate makes the owner very happy, while the hunting state makes the owner quite worried. We can mark each state in Figure 18.2 with a reward score and produce a Markov reward process as in Figure 18.3. Note that what the owner observes may be a chain of states, say "hungry → hunting → hungry → affectionate → sleeping". Then the overall reward, or *value*, will be $(-2)+(-5)+(-2)+(+5)+0 = -4$ along the chain. However, since then, states are observed at different time steps, we need to introduce the notion of *discount* to balance the importance of immediate rewards and long-term rewards when making decisions. For instance, if we take a discount rate of $\gamma = 0.9$, the above value will be $(-2)+(-5)0.9+(-2)0.9^2+(+5)0.9^3+0\cdot0.9^4 = -4.475$.

We can also assign values to each state by setting a system of equations following the one-step process. Since sleeping is a terminal state with zero reward, its value has to be zero. Denote the values of the other states, hunting, hungry, affectionate, relaxed, and confused,

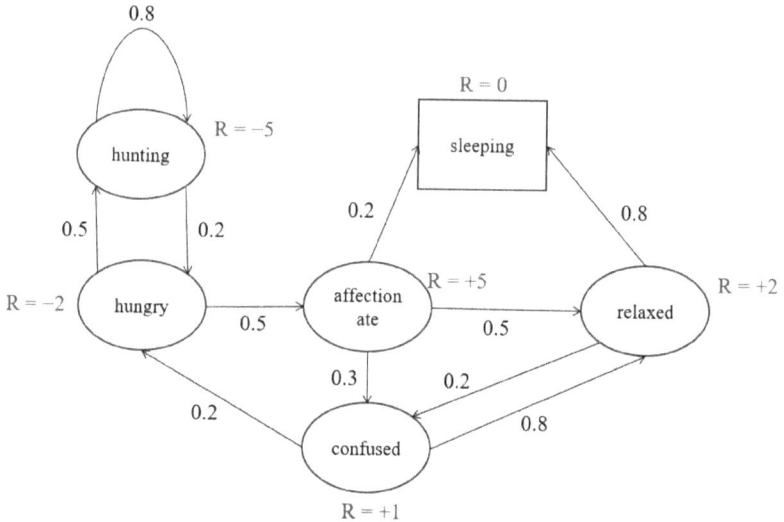

Fig. 18.3. The same Markov chain with rewards on each state.

as $x_1, \ldots, x_5$, respectively. We have

$$x_1 = -5 + 0.8\gamma x_1 + 0.2\gamma x_2, \tag{18.1}$$

$$x_2 = -2 + 0.5\gamma x_1 + 0.5\gamma x_3, \tag{18.2}$$

$$x_3 = 5 + 0.3\gamma x_5 + 0.5\gamma x_4, \tag{18.3}$$

$$x_4 = 2 + 0.2\gamma x_5, \tag{18.4}$$

$$x_5 = 1 + 0.2\gamma x_2 + 0.8\gamma x_4. \tag{18.5}$$

For $\gamma = 0.9$, we have approximately $x_1 = -24.4$, $x_2 = -10.2$, $x_3 = 6.1$, $x_4 = 2.1$, and $x_5 = 0.7$. This is a very simple example of the well-known Bellman equation.

On top of the Markov reward process, an MDP is an environment with a set of actions. For instance, instead of the transition probabilities, if a cat is at the "relaxed" state, the action of "talking" will take it to the "confused" state, while the action of "stroking" will take it to the "sleeping" state. In reinforcement learning, we need to determine value as a function of both the state and the action. Once we are clear about the value, we can take the best action accordingly.

18.3 Deep Q-Learning

When trying to learning the value, there are two important concepts: exploring and exploiting. Exploring refers to the process of trying out new and non-greedy actions to gather more information about the environment, while exploiting refers to choosing best actions according to the information that has already been gathered.

As an example, Q-learning (Watkins and Dayan, 1992) is one of the most common algorithms for RL. In Q-learning, we learn the value Q as a function of the state S and the action A. We randomly initialize $Q(S, A)$ for all possible values of states and actions and then update the values iteratively. The principle for updating $Q(S, A)$ is actually quite similar to the one-step process used to formulate the system of equations in the Markov reward process above. At each state S, we take an action following the policy derived from Q. A possible policy is the ϵ-greedy method, which takes the best action $\arg \max_a Q(S, a)$ (exploiting) with probability $(1-\epsilon)$ and takes a random action with probability ϵ (exploring). After the action is taken, a reward R is observed, which is used to update $Q(S, A)$. To be specific, Q-learning updates Q according to

$$Q(S, A) \leftarrow Q(S, A) + \alpha \left(R + \gamma \max_a Q(S', a) - Q(S, A) \right). \quad (18.6)$$

Here, R is the reward received after taking action A in state S and γ is the discount factor, similar to R and γ in the MDP in the previous section. S' is the state reached after taking A.

In the realm of deep learning, the power of deep neural networks in representing complex and high-dimensional data has been widely demonstrated (as discussed in the previous chapters of this book). In the context of reinforcement learning, the use of deep neural networks for representing value functions or state-action pairs has proven to be highly effective. By leveraging neural networks, reinforcement learning algorithms can capture intricate relationships between actions and rewards, enabling them to learn more sophisticated strategies and generalize to new situations more efficiently compared to traditional methods. A fundamental example of utilizing deep neural networks in reinforcement learning is the method of deep Q-learning (Mnih *et al.*, 2013, 2015), which is a "deep" version of Q-learning.

In deep Q-learning, a neural network $Q(s, a; \boldsymbol{\theta})$ is used to estimate the Q values given the state-action pairs (s, a).

In the earliest version of deep Q-learning (Mnih *et al.*, 2013), which was designed for playing Atari games, the loss function at the ith step is

$$L_i(\boldsymbol{\theta}_i) = \mathbb{E}\left[\frac{1}{2}\left(r + \gamma \max_{a'} Q(s', a'; \boldsymbol{\theta}_{i-1}) - Q(s, a; \boldsymbol{\theta}_i)\right)^2\right]. \quad (18.7)$$

Note that unlike supervised learning, the target value in the above expression, $r + \gamma \max'_a Q(s', a'; \boldsymbol{\theta}_{i-1})$, depends on network parameters. Differentiating the loss function yields

$$\nabla L_i(\boldsymbol{\theta}_i) = \mathbb{E}\left[\left(r + \gamma \max_a' Q(s', a'; \boldsymbol{\theta}_{i-1}) - Q(s, a; \boldsymbol{\theta}_i)\right)\nabla Q(s, a; \boldsymbol{\theta}_i)\right].$$

$$(18.8)$$

We can see that updating parameters according to (18.8) in a stochastic gradient algorithm is very similar to (18.6).

18.4 Policy Gradients

Deep Q-learning is a typical value-based reinforcement learning method since it involves learning the value of each state and action. A different class of models are policy-based, where the policy — a function that maps states to actions — is directly modeled.

A classical such method is REINFORCE (Williams, 1992). In this method, a policy π is typically parameterized, say by $\boldsymbol{\theta}$, which are updated to maximize an expected value via gradient ascent.

Suppose $s_0, a_0, r_0, s_1, a_1, r_1, \ldots, s_T, a_T, r_T$, a sequence of states, actions, and rewards are generated following the current policy $\pi_{\boldsymbol{\theta}}(A|S)$, which can be understood as the distribution of action A given state S. For each $t = 0, \ldots, T - 1$, we can calculate a return $G_t = \sum_{k=t+1}^{T} \gamma^{k-t-1} r_k$, where γ is the discount rate. Denote $J(\boldsymbol{\theta})$ to be the expected value for $\pi_{\boldsymbol{\theta}}$. Then, according to an important result known as the policy gradient theorem,

$$\nabla J(\boldsymbol{\theta}) = \mathbb{E}_{\pi_{\boldsymbol{\theta}}}\left[\gamma^t G_t \frac{\nabla \pi_{\boldsymbol{\theta}}(a_t|s_t)}{\pi_{\boldsymbol{\theta}}(a_t|s_t)}\right] = \mathbb{E}_{\pi_{\boldsymbol{\theta}}}\left[\gamma^t G_t \nabla \log \pi_{\boldsymbol{\theta}}(a_t|s_t)\right].$$

$$(18.9)$$

Hence, to maximize the expected value $J(\boldsymbol{\theta})$, we perform gradient ascent with

$$\boldsymbol{\theta} \leftarrow \boldsymbol{\theta} + \alpha \sum_{t=0}^{T-1} \sum_{k=t+1}^{T} \gamma^{k-1} \mathbb{E}_{\pi_{\boldsymbol{\theta}}} \left[r_k \nabla \log \pi_{\boldsymbol{\theta}}(a_t|s_t) \right]. \tag{18.10}$$

We can loop the sequence generation above to keep updating $\boldsymbol{\theta}$.

Although the classical REINFORCE is intuitive, its training can suffer from instability and slow convergence. One solution is to reduce the variance by introducing a baseline $b(s_t)$ and replace G_t in (18.9) with $G_t - b(s_t)$.

In the era of deep learning, numerous methods have been developed to enhance this approach. Among these, the actor-critic methods stand out as a particularly significant advancement.

18.5   Actor-Critic Methods

As we have seen, the primary distinction between DQL and REIN-FORCE lies in their learning focuses: DQL learns the value function, whereas REINFORCE directly optimizes the policy. Actor-critic methods unite both approaches by simultaneously learning the value function and optimizing the policy. Specifically, the actor is a parameterized policy function $\pi_{\boldsymbol{\theta}}(A|S)$ and the critic is a parameterized value function $v_{\mathbf{w}}(S)$. The actor is updated with the gradient in (18.9) replaced with (more generally)

$$\mathbb{E}_{\pi_{\boldsymbol{\theta}}} \left[\delta_t \nabla \log \pi_{\boldsymbol{\theta}}(a_t|s_t) \right]. \tag{18.11}$$

Here, δ_t can be taken to be the advantage function $\mathcal{A}(s_t, a_t)$, which measures the difference between an action-value function $Q(s_t, a_t)$ and a state-value function $V(s_t)$. For instance, we can take $Q(s_t, a_t) = r_t + \gamma v_{\mathbf{w}}(s_{t+1})$ and $V(s_t) = v_{\mathbf{w}}(s_t)$, so that $\delta_t = r_t + \gamma v_{\mathbf{w}}(s_{t+1}) - v_{\mathbf{w}}(s_t)$. In this case, the critic is updated by minimizing the MSE of δ_t. Treating $r_t + \gamma v_{\mathbf{w}}(s_{t+1})$ as a target, the gradient of $\frac{1}{2}(r_t + \gamma v_{\mathbf{w}}(s_{t+1}) - v_{\mathbf{w}}(s_t))^2$ is

$$-(r_t + \gamma v_{\mathbf{w}}(s_{t+1}) - v_{\mathbf{w}}(s_t)) \nabla v_{\mathbf{w}}(s_t), \tag{18.12}$$

which is similar to (18.8).

We introduce two well-known methods within this family. The first is the trusted region policy optimization (TRPO) method (Schulman *et al.*, 2015). In TRPO, a trusted region is employed so that the new policy does not differ too much from the old policy. Actually, the way TRPO builds the trusted region is to set a threshold for the KL divergence (which has appeared so many times in this book!) between the policies.

However, TRPO adopts a second-order optimization and is computationally expensive. This has restricted its adoption in large-scale problems. To this end, Schulman *et al.* (2017) proposed the proximal policy optimization (PPO) method. Instead of imposing a hard KL-divergence constraint with complex computations, PPO uses a clipped version of the objective function. If the ratio between the new policy (after gradient update) and the current policy is outside a certain range $[1-\epsilon, 1+\epsilon]$, it will be penalized in the objective function. PPO is designed for easier implementation, but its performance is usually similar to TRPO.

18.6   AlphaGo

Deep reinforcement learning is currently a hot field in the area of artificial intelligence and machine learning and there are a large number of state-of-the-art applications. In this section, we briefly introduce the algorithm behind AlphaGo, the program developed by DeepMind for playing the board game Go, whose victory against Lee Sedol was seen as a major milestone in the development of artificial intelligence and the field of deep reinforcement learning.

AlphaGo was developed in 2016 (Silver *et al.*, 2016). It used a combination of deep neural networks and Monte Carlo tree search (MCTS) algorithms to learn from large datasets of human and computer-generated games and to simulate future moves and outcomes to select the best possible move in each situation. In 2017, DeepMind announced AlphaGo Zero (Silver *et al.*, 2017b), which was a significant improvement over AlphaGo. AlphaGo Zero was designed to learn entirely from scratch, without any human data or intervention. It used a similar architecture to AlphaGo, with deep neural networks and Monte Carlo tree search, but was trained solely by playing against itself millions of times. AlphaGo Zero was able to surpass the performance of the original AlphaGo and achieve super-human levels of play within just a few days of training. It also demonstrated a more efficient and effective approach to deep reinforcement learning, showing that it is possible to achieve highly sophisticated AI without relying on large amounts of human data. Like AlphaGo Zero, AlphaZero (Silver *et al.*, 2017a) is a reinforcement learning algorithm that learns entirely from scratch and requires no human knowledge or data to learn. However, AlphaZero was designed to be a more general-purpose algorithm that can play not only Go but also other board games such as chess and shogi.

We mainly focus on AlphaGo Zero to briefly discuss the main techniques. In AlphaGo Zero, there is a neural network that represents both the policy and the value. The MCTS, which we elaborate later, is used for self-play and generation of training data for updating the neural network. Meanwhile, the MCTS is also guided by the neural network. Given $\mathbf{s}$, the state of the board, the deep neural network $f_{\boldsymbol{\theta}}$ (where $\boldsymbol{\theta}$ describes the parameters) produces

$$(\mathbf{p}, v) = f_{\boldsymbol{\theta}}(\mathbf{s}). \tag{18.13}$$

Here, $\mathbf{p}$ is a vector that describes the move probability: $p_a = \mathbb{P}(a|\mathbf{s})$ for any action a (i.e., $\mathbf{p}$ is the policy). v is a scalar that represents the probability of the current player winning given the state $\mathbf{s}$ of the board (i.e., v is the value).

Specifically to the game of Go, the state $\mathbf{s}$ is given by a $19 \times 19 \times 17$ binary tensor. Here, 19×19 is the size of the board, and $17 = 8+8+1$ keeps 8 black history, 8 white history, and 1 for which player to play, which is due to special rules of the competition. $f_{\boldsymbol{\theta}}$ is a ReLU CNN. Given the state of the art at that time, the CNN contains residual layers and batch normalization. As to the output, $\mathbf{p}$ is a vector of size $19 \times 19 + 1$, where the additional "1" is due to the rule that a player may choose to pass. On the other hand, $v \in [-1, 1]$ is a single scalar.

Assume we already have a regime for generating self-play, which generates a sequence of states of the board $\mathbf{s}_1, \ldots, \mathbf{s}_T$, until a winner is determined. We have a search probability $\boldsymbol{\pi}_t$ at each given state $\mathbf{s}_t$, which has the same size $19 \times 19 + 1$ as $\mathbf{p}$. There will also be a result z from self-play, indicating the winner of the game. These will be used for training the neural network with the loss

$$\ell = (z - v)^2 - \boldsymbol{\pi}^\top \log \mathbf{p} + c \, \|\boldsymbol{\theta}\|^2, \qquad (18.14)$$

where the last term is a regularization term which prevents overfitting. In AlphaGo Zero, a checkpoint is produced after each 1,000 training steps to evaluate the performance of the current network. The evaluation is done by comparing the current network against the previous best network in a set of self-play games. If the current network achieves a higher win rate, it replaces the previous best network as the new benchmark. On the other hand, in AlphaZero, the most recent network is always used during self-play, and there is no explicit checkpoint evaluation process like in AlphaGo Zero. However, the network is periodically updated based on the results of the self-play games. Specifically, the network is updated using a weighted average of the latest network and a previous version of the network, where the weight is determined by the outcome of the self-play games. If the latest network wins a certain threshold percentage of the games, it replaces the previous version as the new network.

Next, we describe the Monte Carto Tree Search (MCTS) algorithm used in AlphaGo Zero, guided by the above neural network. As its name suggests, MCTS is an algorithm that builds a search tree of possible moves and outcomes to measure the value of each move. There are two important terms that guide the moves, namely the (average) action value, denoted as $Q(s, a)$, and the upper confidence bound (UCB), denoted as $U(s, a)$. Here, s denotes a node of the tree, and a is chosen from all legal actions. A move intends to maximize the sum of the action value and the UCB. Specifically, the action value of a move is the average outcome from simulations involving this move. It reflects the current estimate of the value of the move. On the other hand, the UCB is a term that decreases when the number of simulations increases. It encourages the algorithm to take new moves. The sum of the action value and the UCB reflects a balance of exploiting and exploring.

The MCTS algorithm starts at the root node of the search tree at time 0. It finishes when it reaches a leaf node at time L. The main steps involved in MCTS are select, expand and evaluate, backup, and play:

(1) **Select**: Given a node s and an action a, we denote the visit count by $N(s, a)$, the total action value by $W(s, a)$, and the prior probability of selecting an action by $P(s, a)$. At each time step $t = 1, \ldots, L-1$, the action a_t is chosen by maximizing $Q(s_t, a) + U(s_t, a)$, where

$$U(s, a) = cP(s, a)\frac{\sqrt{\sum_b N(s, b)}}{1 + N(s, a)}. \tag{18.15}$$

Here, c is a hyperparameter that balances exploiting and exploring.

(2) **Expand and evaluate**: The leaf node s_L is evaluated by the neural network f_θ to produce the move probability $\mathbf{p}$. Each action a corresponds then to an initialization of

$$\{N(s_L, a) = 0, W(s_L, a) = 0, Q(s_L, a) = 0, P(s_L, a) = p_a\},$$

where p_a is the entry in $\mathbf{p}$ corresponding to a.

(3) **Backup**: The values of N, W, Q are updated backward for time steps $t \leq L$. The updated values are $N(s_t, a_t) \leftarrow N(s_t, a_t) + 1$, $W(s_t, a_t) \leftarrow W(s_t, a_t) + v$ where v is the value output of f_θ and $Q(s_t, a_t) \leftarrow W(s_t, a_t)/N(s_t, a_t)$.

(4) **Play**: After search, a move a is selected at the root s_0, with

$$\pi(a|s_0) = \frac{N(s_0, a)^{1/\tau}}{\sum_b N(s_0, b)^{1/\tau}}. \tag{18.16}$$

Conclusion

This book started as a set of lecture notes from the deep learning course that we co-taught. While those notes offered a blueprint, the work of transforming them into a book has been a journey filled with both excitement and discovery, as new ideas in deep learning continually emerged and reshaped our own perspectives and the structure of the course every year.

As we started the project, it was challenging to predict the precise scope and coverage of this book. Deep learning is a field that advances with remarkable speed, and as we wrote, new concepts and methods appeared, influencing what we considered essential to include. For instance, as we started, diffusion models have not come to stage yet.

Unlike what other books in deep learning might do, we do not spend many chapters on foundational topics in basic machine learning that students are expected to grasp from prerequisite courses. Instead, we concentrated on ideas that play crucial roles such as convolutional neural networks, recurrent neural networks, and generative models. These form the core upon which many newer developments build, making them essential to any deep learning curriculum.

Beyond these foundational architectures, we wanted to delve into topics that could foster deeper insights for students ready to explore the inner workings of deep learning models at a more theoretical level. Two ideas stood out: the Information Bottleneck (IB) and the Neural Tangent Kernel (NTK). They were proposed from different interests, yet as we explored them more closely, intriguing connections emerged. Both the IB and NTK frameworks shed light on how neural networks propagate and generalize. Moreover, NTK provides a tractable tool that makes it possible to calculate information theoretical quantities. Indeed, we do find works that bring together both approaches. For instance, Shwartz-Ziv and Alemi (2020); Adnan *et al.* (2022) both used NTK to simplify the calculation of mutual information in order to study the generalizability of deep neural networks.

While these advanced topics push students to a deeper understanding, we also recognize that the field will continue to grow. New ideas will undoubtedly arise, and with them, the boundaries of this book will inevitably shift. Staying up-to-date in such a rapidly evolving field requires an open mind and a willingness to continually explore and adapt. We hope that this book will serve not only as a foundation but also as a springboard for students and practitioners, encouraging them to keep pace with the field and to contribute to its development. The journey of deep learning is ongoing, and we are excited to have been a part of this ever-evolving story.

Bibliography

Abdallah, S. A. and Plumbley, M. D. (2012). Predictive information rate in discrete-time Gaussian processes. arXiv:1206.0304 [stat.ML].

Adnan, M., Ioannou, Y., Tsai, C.-Y., Galloway, A., Tizhoosh, H. R., and Taylor, G. W. (2022). Monitoring shortcut learning using mutual information. *arXiv preprint* arXiv:2206.13034.

Alemi, A., Poole, B., Fischer, I., Dillon, J., Saurous, R. A., and Murphy, K. (2018). Fixing a broken ELBO. In *International Conference on Machine Learning*. PMLR, pp. 159–168.

Arjovsky, M. and Bottou, L. (2017). Towards principled methods for training generative adversarial networks. In *International Conference on Learning Representations*.

Arjovsky, M., Chintala, S., and Bottou, L. (2017). Wasserstein generative adversarial networks. In *International Conference on Machine Learning* (pp. 214–223). PMLR.

Arora, S., Du, S., Hu, W., Li, Z., and Wang, R. (2019a). Fine-grained analysis of optimization and generalization for overparameterized two-layer neural networks. In *International Conference on Machine Learning*. PMLR, pp. 322–332.

Arora, S., Du, S. S., Hu, W., Li, Z., Salakhutdinov, R. R., and Wang, R. (2019b). On exact computation with an infinitely wide neural net. In *Advances in Neural Information Processing Systems*, Vol. 32.

Babu, G. J., Banks, D., Cho, H., Han, D., Sang, H., and Wang, S. (2021). A statistician teaches deep learning. *Journal of Statistical Theory and Practice* **15**, 1–23.

Belghazi, M. I., Baratin, A., Rajeswar, S., Ozair, S., Bengio, Y., Courville, A., and Hjelm, R. D. (2018). Mutual information neural estimation. In *Proceedings of the 35th International Conference on Machine Learning*. https://arxiv.org/abs/1801.04062.

Bishop, C. M. and Bishop, H. (2023). *Deep Learning: Foundations and Concepts*. Springer Nature.

Bishop, C. M. and Nasrabadi, N. M. (2006). *Pattern Recognition and Machine Learning* (Vol. 4, No. 4, p. 738). New York: Springer.

Bottou, L., Curtis, F. E., and Nocedal, J. (2018). Optimization methods for large-scale machine learning. *SIAM Review* **60**(2), 223–311.

Boyd, S. P. and Vandenberghe, L. (2004). *Convex Optimization*. Cambridge: Cambridge University Press.

Brown, T. B., Mann, B., Ryder, N., Subbiah, M., Kaplan, J., Dhariwal, P., Neelakantan, A., Shyam, P., Sastry, G., Askell, A., Agarwal, S., Herbert-Voss, A., Krueger, G., Henighan, T., Child, R., Ramesh, A., Ziegler, D. M., Wu, J., Winter, C., Hesse, C., Chen, M., Sigler, E., Litwin, M., Gray, S., Chess, B., Clark, J., Berner, C., McCandlish, S., Radford, A., Sutskever, I., and Amodei, D. (2020). Language models are few-shot learners. In *Proceedings of the 34th International Conference on Neural Information Processing Systems*, NIPS'20. Curran Associates Inc., Red Hook, NY, USA.

Cao, K., Rong, Y., Li, C., Tang, X., and Loy, C. C. (2018). Pose-robust face recognition via deep residual equivariant mapping. In *Proceedings of the IEEE Conference on Computer Vision and Pattern Recognition*, pp. 5187–5196.

Chen, L.-C., Papandreou, G., Kokkinos, I., Murphy, K., and Yuille, A. L. (2017a). Deeplab: Semantic image segmentation with deep convolutional nets, atrous convolution, and fully connected CRFs. *IEEE Transactions on Pattern Analysis and Machine Intelligence* **40**(4), 834–848.

Chen, L.-C., Papandreou, G., Schroff, F., and Adam, H. (2017b). Rethinking atrous convolution for semantic image segmentation. *arXiv preprint* arXiv:1706.05587.

Cho, A., Kim, G. C., Karpekov, A., Helbling, A., Wang, Z. J., Lee, S., Hoover, B., and Chau, D. H. (2024). Transformer explainer: Interactive learning of text-generative models. arXiv:2408.04619 [cs.LG], https://arxiv.org/abs/2408.04619.

Cover, T. M. and Thomas, J. (1991). *Elements of Information Theory*. Hoboken, N.J.: Wiley.

Daubechies, I. (1992). *Ten Lectures on Wavelets*. Philadelphia, Pennsylvania: SIAM.

Devlin, J., Chang, M.-W., Lee, K., and Toutanova, K. (2019). Bert: Pre-training of deep bidirectional transformers for language understanding. *ArXiv* abs/1810.04805.

Dhariwal, P. and Nichol, A. (2021). Diffusion models beat gans on image synthesis. In *Advances in Neural Information Processing Systems*, Vol. 34, pp. 8780–8794.

Donsker, M. and Varadhan, S. (1983). Asymptotic evaluation of certain markov process expectations for large time. IV. *Communications on Pure and Applied Mathematics* **36**(2), 183–212.

Dosovitskiy, A., Beyer, L., Kolesnikov, A., Weissenborn, D., Zhai, X., Unterthiner, T., Dehghani, M., Minderer, M., Heigold, G., Gelly, S., Uszkoreit, J., and Houlsby, N. (2021). An image is worth 16x16 words: Transformers for image recognition at scale, in *International Conference on Learning Representations*, https://openreview.net/forum?id=YicbFdNTTy.

Dubnov, S., Chen, K., and Huang, K. (2022). Deep music information dynamics. *Journal of Creative Music Systems* **1**(1). DOI: https://doi.org/10.5920/jcms.894.

Elizalde, B., Deshmukh, S., Ismail, M. A., and Wang, H. (2023). Clap learning audio concepts from natural language supervision. In *ICASSP 2023-2023 IEEE International Conference on Acoustics, Speech and Signal Processing (ICASSP)*, pp. 1–5. https://api.semanticscholar.org/CorpusID:249605738.

Frey, B. J., Dayan, P., and Hinton, G. E. (1997). *A Simple Algorithm That Discovers Efficient Perceptual Codes*. Cambridge University Press, USA, pp. 296–315.

Frey, B. J. and Hinton, G. E. (1996). Free energy coding. In *Proceedings of the Conference on Data Compression, DCC'96*. IEEE Computer Society, USA, p. 73.

Ganin, Y., Ustinova, E., Ajakan, H., Germain, P., Larochelle, H., Laviolette, F., Marchand, M., and Lempitsky, V. (2016). Domain-adversarial training of neural networks. *The Journal of Machine Learning Research* **17**(1), 2096–2030.

Geng, H., Hu, Y., and Huang, H. (2020). Monaural singing voice and accompaniment separation based on gated nested U-Net architecture. *Symmetry* **12**(6), 1051.

Ghadimi, S. and Lan, G. (2013). Stochastic first-and zeroth-order methods for nonconvex stochastic programming. *SIAM Journal on Optimization* **23**(4), 2341–2368.

Girshick, R. (2015). Fast R-CNN. In *Proceedings of the IEEE International Conference on Computer Vision*, pp. 1440–1448.

Girshick, R., Donahue, J., Darrell, T., and Malik, J. (2015). Region-based convolutional networks for accurate object detection and segmentation. *IEEE Transactions on Pattern Analysis and Machine Intelligence* **38**(1), 142–158.

Glorot, X. and Bengio, Y. (2010). Understanding the difficulty of training deep feedforward neural networks. In *Proceedings of the Thirteenth International Conference on Artificial Intelligence and Statistics*. JMLR Workshop and Conference Proceedings, pp. 249–256.

Goodfellow, I., Bengio, Y., and Courville, A. (2016). *Deep Learning*. Cambridge, Massachusetts: MIT Press. http://www.deeplearningbook.org.

Goodfellow, I., Pouget-Abadie, J., Mirza, M., Xu, B., Warde-Farley, D., Ozair, S., Courville, A., and Bengio, Y. (2014). Generative adversarial nets. In *Advances in Neural Information Processing Systems*, Vol. 27.

Goodfellow, I., Shlens, J., and Szegedy, C. (2015). Explaining and harnessing adversarial examples. In *International Conference on Learning Representations*.

Gretton, A., Borgwardt, K. M., Rasch, M. J., Schölkopf, B., and Smola, A. (2012). A kernel two-sample test. *The Journal of Machine Learning Research* **13**(1), 723–773.

Gulrajani, I., Ahmed, F., Arjovsky, M., Dumoulin, V., and Courville, A. C. (2017). Improved training of wasserstein gans. In *Advances in Neural Information Processing Systems*.

Han, Z., Gao, C., Liu, J., Zhang, J., and Zhang, S. Q. (2024). Parameter-efficient fine-tuning for large models: A comprehensive survey. arXiv:2403.14608 [cs.LG], https://arxiv.org/abs/2403.14608.

Harris, Z. S. (1954). Distributional structure. *Word* **10**(2–3), 146–162.

He, K., Zhang, X., Ren, S., and Sun, J. (2016). Deep residual learning for image recognition. In *Proceedings of the IEEE Conference on Computer Vision and Pattern Recognition*, pp. 770–778.

Hendrycks, D. and Gimpel, K. (2016). Gaussian error linear units (gelus). *arXiv preprint* arXiv:1606.08415.

Herremans, D. and Chuan, C.-H. (2017). Modeling musical context with word2vec. *arXiv preprint* arXiv:1706.09088.

Hofmann, T., Schölkopf, B., and Smola, A. J. (2008). Kernel methods in machine learning.

Howard, J. and Gugger, S. (2020). *Deep Learning for Coders with fastai and PyTorch*. Sebastopol, California: O'Reilly Media.

Hu, E. J., Shen, Y., Wallis, P., Allen-Zhu, Z., Li, Y., Wang, L., Wang, W., and Chen, W.-L. (2021). Lora: Low-rank adaptation of large language models. *arXiv preprint* arXiv:2106.09685.

Huang, A., Dinculescu, M., Vaswani, A., and Eck, D. (2018a). Visualizing music self-attention. In *NIPS Workshop on Interpretability and Robustness in Audio, Speech, and Language*. https://openreview.net/pdf?id=ryfxVNEajm.

Huang, C.-W., Krueger, D., Lacoste, A., and Courville, A. (2018b). Neural autoregressive flows. In *International Conference on Machine Learning*. PMLR, pp. 2078–2087.

Ioffe, S. and Szegedy, C. (2015). Batch normalization: Accelerating deep network training by reducing internal covariate shift. In *International Conference on Machine Learning*. PMLR, pp. 448–456.

Jacot, A., Gabriel, F., and Hongler, C. (2018). Neural tangent kernel: Convergence and generalization in neural networks. In *Advances in Neural Information Processing Systems*, Vol. 31.

Jourabloo, A., Liu, Y., and Liu, X. (2018). Face de-spoofing: Anti-spoofing via noise modeling. In *Proceedings of the European Conference on Computer Vision (ECCV)*, pp. 290–306.

Jurafsky, D. and Martin, J. H. (2022). *Speech and Language Processing*, 3rd edn. https://web.stanford.edu/~jurafsky/slp3/.

Kae, A., Sohn, K., Lee, H., and Learned-Miller, E. (2013). Augmenting CRFs with boltzmann machine shape priors for image labeling. In *Proceedings of the IEEE Conference on Computer Vision and Pattern Recognition*, pp. 2019–2026.

Ketkar, N. and Santana, E. (2017). *Deep Learning with Python*, Vol. 1. Berlin, Heidelberg, Germany: Springer.

Kingma, D. P. and Ba, J. (2015). Adam: A method for stochastic optimization. In *International Conference on Learning Representations*.

Kingma, D. P. and Welling, M. (2014). Auto-encoding variational bayes.

Kirzhevsky, A., Sutskever, I., and Hinton, G. E. (2012). Imagenet classification with deep convolutional neural networks. In *Advances in Neural Information Processing Systems*, Vol. 25, pp. 1097–1105.

Kobyzev, I., Prince, S. J., and Brubaker, M. A. (2020). Normalizing flows: An introduction and review of current methods. *IEEE Transactions on Pattern Analysis and Machine Intelligence* **43**(11), 3964–3979.

Lei, T., Bai, J., Brahma, S., Ainslie, J., Lee, K., Zhou, Y., Du, N., Zhao, V. Y., Wu, Y., and Li, B. E. A. (2023). Conditional adapters: Parameter-efficient transfer learning with fast inference. *arXiv preprint* arXiv:2304.04947.

Li, H., Xu, Z., Taylor, G., Studer, C., and Goldstein, T. (2018). Visualizing the loss landscape of neural nets. In *Advances in Neural Information Processing Systems*, Vol. 31.

Lundberg, S. M. and Lee, S.-I. (2017). A unified approach to interpreting model predictions. In *31st Conference on Neural Information Processing Systems*.

Mallat, S. (2012). Group invariant scattering. *Communications on Pure and Applied Mathematics* **65**(10), 1331–1398.

Mienye, I. D., Swart, T. G., and Obaido, G. (2024). Recurrent neural networks: A comprehensive review of architectures, variants, and applications. *Information* **15**(9). https://www.mdpi.com/2078-2489/15/9/517.

Mikolov, T., Chen, K., Corrado, G., and Dean, J. (2013). Efficient estimation of word representations in vector space. *arXiv preprint* arXiv:1301.3781.

Mnih, V., Kavukcuoglu, K., Silver, D., Graves, A., Antonoglou, I., Wierstra, D., and Riedmiller, M. (2013). Playing atari with deep reinforcement learning. In *Advances in Neural Information Processing Systems*, pp. 2672–2680.

Mnih, V., Kavukcuoglu, K., Silver, D., Rusu, A. A., Veness, J., Bellemare, M. G., Graves, A., Riedmiller, M., Fidjeland, A. K., Ostrovski, G., *et al.* (2015). Human-level control through deep reinforcement learning. *Nature* **518**(7540), 529–533.

Murray, N., Marchesotti, L., and Perronnin, F. (2012). AVA: A large-scale database for aesthetic visual analysis. In *2012 IEEE Conference on Computer Vision and Pattern Recognition*. IEEE, pp. 2408–2415.

Novak, R., Xiao, L., Hron, J., Lee, J., Alemi, A. A., Sohl-Dickstein, J., and Schoenholz, S. S. (2020). Neural tangents: Fast and easy infinite neural networks in python. In *International Conference on Learning Representations*. https://openreview.net/forum?id=SklD9yrFPS.

Nowozin, S., Cseke, B., and Tomioka, R. (2016). f-GAN: Training generative neural samplers using variational divergence minimization. In *Advances in Neural Information Processing Systems*, Vol. 29.

Papamakarios, G., Nalisnick, E., Rezende, D. J., Mohamed, S., and Lakshminarayanan, B. (2021). Normalizing flows for probabilistic modeling and inference. *The Journal of Machine Learning Research* **22**(1), 2617–2680.

Pennington, J., Socher, R., and Manning, C. D. (2014). Glove: Global vectors for word representation. In *Proceedings of the 2014 Conference on Empirical Methods in Natural Language Processing (EMNLP)*, pp. 1532–1543.

Peyré, G., Cuturi, M., *et al.* (2019). Computational optimal transport: With applications to data science. *Foundations and Trends in Machine Learning* **11**(5–6), 355–607.

Piasini, F. A. L. J., E. and Gold, J. (2021). Embo: A python package for empirical data analysis using the information bottleneck. *Journal of Open Research Software* **9**(1), 1–7.

Poth, C., Sterz, H., Paul, I., Purkayastha, S., Engländer, L., Imhof, T., Vulic, I., Ruder, S., Gurevych, I., and Pfeiffer, J. (2023). Adapters: A unified library for parameter-efficient and modular transfer learning.

Radford, A., Narasimhan, K., Salimans, T., and Sutskever, I. (2018). Improving language understanding by generative pre-training. OpenAI Technical Report. https://cdn.openai.com/research-covers/language-unsupervised/language_understanding_paper.pdf.

Radford, A., Kim, J. W., Hallacy, C., Ramesh, A., Goh, G., Agarwal, S., Sastry, G., Askell, A., Mishkin, P., Clark, J., Krueger, G., and Sutskever, I. (2021). Learning transferable visual models from natural language supervision. In M. Meila and T. Zhang (eds.), *Proceedings of the 38th International Conference on Machine Learning*, *Proceedings of Machine Learning Research*, Vol. 139. PMLR, pp. 8748–8763. https://proceedings.mlr.press/v139/radford21a.html.

Radford, A., Metz, L., and Chintala, S. (2016). Unsupervised representation learning with deep convolutional generative adversarial networks. In *International Conference on Learning Representations*.

Radford, A., Wu, J., Child, R., Luan, D., Amodei, D., and Sutskever, I. (2019). *Language Models are Unsupervised Multitask Learners.*

Redmon, J., Divvala, S., Girshick, R., and Farhadi, A. (2016). You only look once: Unified, real-time object detection. In *Proceedings of the IEEE Conference on Computer Vision and Pattern Recognition*, pp. 779–788.

Redmon, J. and Farhadi, A. (2016). Yolo9000: Better, faster, stronger, 2016. **394**. *arXiv preprint* arXiv:1612.08242.

Redmon, J. and Farhadi, A. (2018). Yolov3: An incremental improvement. *arXiv preprint* arXiv:1804.02767.

Ren, S., He, K., Girshick, R., and Sun, J. (2015). Faster R-CNN: Towards real-time object detection with region proposal networks. In *Advances in Neural Information Processing Systems*, Vol. 28.

Ronneberger, O., Fischer, P., and Brox, T. (2015). U-Net: Convolutional networks for biomedical image segmentation. In *Medical Image Computing and Computer-Assisted Intervention–MICCAI 2015: 18th International Conference, Munich, Germany, October 5–9, 2015, Proceedings, Part III 18*. Springer, pp. 234–241.

Ruthotto, L. and Haber, E. (2021). An introduction to deep generative modeling. *GAMM-Mitteilungen* **44**(2), e202100008.

Salimans, T., Goodfellow, I., Zaremba, W., Cheung, V., Radford, A., and Chen, X. (2016). Improved techniques for training GANs. In *Advances in Neural Information Processing Systems*, Vol. 29.

Saxe, A. M., Bansal, Y., Dapello, J., Advani, M., Kolchinsky, A., Tracey, B. D., and Cox, D. D. (2018). On the information bottleneck theory of deep learning. In *International Conference on Learning Representations*. https://openreview.net/forum?id=ry_WPG-A-.

Schulman, J., Levine, S., Abbeel, P., Jordan, M., and Moritz, P. (2015). Trust region policy optimization. In *International Conference on Machine Learning*. PMLR, pp. 1889–1897.

Schulman, J., Wolski, F., Dhariwal, P., Radford, A., and Klimov, O. (2017). Proximal policy optimization algorithms. *arXiv preprint* arXiv:1707.06347.

Sengupta, S., Chen, J.-C., Castillo, C., Patel, V. M., Chellappa, R., and Jacobs, D. W. (2016). Frontal to profile face verification in the wild. In *2016 IEEE Winter Conference on Applications of Computer Vision (WACV)*. IEEE, pp. 1–9.

Shalev-Shwartz, S. and Ben-David, S. (2014). *Understanding Machine Learning: From Theory to Algorithms*. Cambridge: Cambridge University Press.

Shaw, P., Uszkoreit, J., and Vaswani, A. (2018). Self-attention with relative position representations. In M. Walker, H. Ji, and A. Stent (eds.), *Proceedings of the 2018 Conference of the North American Chapter of the Association for Computational Linguistics: Human Language Technologies, Volume 2 (Short Papers)*. Association for Computational Linguistics, New Orleans, Louisiana, pp. 464–468. DOI: 10.18653/v1/N18-2074, https://aclanthology.org/N18-2074.

Shwartz-Ziv, R. and Alemi, A. A. (2020). Information in infinite ensembles of infinitely-wide neural networks. In *Symposium on Advances in Approximate Bayesian Inference*. PMLR, pp. 1–17.

Silver, D. (2015). Introduction to reinforcement learning with David Silver. *DeepMind x UCL*.

Silver, D., Huang, A., Maddison, C. J., Guez, A., Sifre, L., Van Den Driessche, G., Schrittwieser, J., Antonoglou, I., Panneershelvam, V., Lanctot, M., *et al.* (2016). Mastering the game of go with deep neural networks and tree search. *Nature* **529**(7587), 484–489.

Silver, D., Hubert, T., Schrittwieser, J., Antonoglou, I., Lai, M., Guez, A., Lanctot, M., Sifre, L., Kumaran, D., Graepel, T., *et al.* (2017a). Mastering chess and shogi by self-play with a general reinforcement learning algorithm. *arXiv preprint* arXiv:1712.01815.

Silver, D., Schrittwieser, J., Simonyan, K., Antonoglou, I., Huang, A., Guez, A., Hubert, T., Baker, L., Lai, M., Bolton, A., *et al.* (2017b). Mastering the game of go without human knowledge. *Nature* **550**(7676), 354–359.

Simon, I. and Oore, S. (2017). Performance RNN: Generating music with expressive timing and dynamics. https://magenta.tensorflow.org/performance-rnn.

Simonyan, K. and Zisserman, A. (2015). Very deep convolutional networks for large-scale image recognition. In *International Conference on Learning Representations*.

Skorski, M., Temperoni, A., and Theobald, M. (2021). Revisiting weight initialization of deep neural networks. In *Asian Conference on Machine Learning*. PMLR, pp. 1192–1207.

Smith, L. N. and Topin, N. (2019). Super-convergence: Very fast training of neural networks using large learning rates. In *Artificial Intelligence and Machine Learning for Multi-Domain Operations Applications*, Vol. 11006. SPIE, pp. 369–386.

Song, Y. and Ermon, S. (2019). Generative modeling by estimating gradients of the data distribution. In *Advances in Neural Information Processing Systems*, Vol. 32.

Su, W., Boyd, S., and Candes, E. (2014). A differential equation for modeling nesterov's accelerated gradient method: Theory and insights. In *Advances in Neural Information Processing Systems*, Vol. 27.

Sutskever, I., Martens, J., Dahl, G., and Hinton, G. (2013). On the importance of initialization and momentum in deep learning. In *International Conference on Machine Learning*. PMLR, pp. 1139–1147.

Sutton, R. S. and Barto, A. G. (2018). *Reinforcement Learning: An Introduction*. Cambridge, Massachusetts: MIT Press.

Szegedy, C., Zaremba, W., Sutskever, I., Bruna, J., Erhan, D., Goodfellow, I., and Fergus, R. (2013). Intriguing properties of neural networks. In *International Conference on Learning Representations*.

Tay, Y., Dehghani, M., Bahri, D., and Metzler, D. (2022). Efficient transformers: A survey. *ACM Computing Surveys* **55**(6), 1–28.

Tipping, M. E. and Bishop, C. M. (1999). Probabilistic principal component analysis. *Journal of the Royal Statistical Society Series B: Statistical Methodology* **61**(3), 611–622.

Tishby, N. and Zaslavsky, N. (2015). Deep learning and the information bottleneck principle. In *Information Theory Workshop (ITW)*, pp. 1–5.

Van Den Oord, A., Kalchbrenner, N., and Kavukcuoglu, K. (2016). Pixel recurrent neural networks. In *International Conference on Machine Learning*. PMLR, pp. 1747–1756.

Vaswani, A., Shazeer, N., Parmar, N., Uszkoreit, J., Jones, L., Gomez, A. N., Kaiser, L., and Polosukhin, I. (2017). Attention is all you need. In *Proceedings of the 31st International Conference on Neural Information Processing Systems, NIPS'17*. Curran Associates Inc., Red Hook, NY, USA, p. 6000–6010.

Villani, C. (2003). *Topics in Optimal Transportation*. Providence, Rhode Island: American Mathematical Society.

Vincent, P. (2011). A connection between score matching and denoising autoencoders. *Neural Computation* **23**(7), 1661–1674.

Wang, Y. (2020). A mathematical introduction to generative adversarial nets (gan). *arXiv preprint* arXiv:2009.00169.

Watkins, C. J. and Dayan, P. (1992). Q-learning. *Machine Learning* **8**, 279–292.

Williams, C. K. and Rasmussen, C. E. (2006). *Gaussian Processes for Machine Learning*, Vol. 2. MIT Press, Cambridge, MA.

Williams, R. J. (1992). Simple statistical gradient-following algorithms for connectionist reinforcement learning. *Machine Learning* **8**, 229–256.

Wu, Y., Chen, K., Zhang, T., Hui, Y., Berg-Kirkpatrick, T., and Dubnov, S. (2022). Large-scale contrastive language-audio pretraining with feature fusion and keyword-to-caption augmentation. In *ICASSP 2023–2023 IEEE International Conference on Acoustics, Speech and Signal Processing (ICASSP)*, pp. 1–5. https://api.semanticscholar. org/CorpusID:253510826.

Wu, Y., Chen, K., Zhang, T., Hui, Y., Nezhurina, M., Berg-Kirkpatrick, T., and Dubnov, S. (2023). Large-scale contrastive language-audio pretraining with feature fusion and keyword-to-caption augmentation. In *IEEE International Conference on Acoustics, Speech and Signal Processing (ICASSP)*. https://arxiv.org/abs/2211.06687.

Yang, D., Yu, J., Wang, H., Wang, W., Weng, C., Zou, Y., and Yu, D. (2023). Diffsound: Discrete diffusion model for text-to-sound generation. *IEEE/ACM Transactions on Audio, Speech, and Language Processing* **31**, 1720–1733. DOI: 10.1109/TASLP.2023.3268730.

Yosinski, J., Clune, J., Bengio, Y., and Lipson, H. (2014). How transferable are features in deep neural networks? In *Advances in Neural Information Processing Systems*, Vol. 27.

Young, H. (1995). *Equity: In Theory and Practice*. Princeton, New Jersey: Princeton University Press.

Zeiler, M. D. and Fergus, R. (2014). Visualizing and understanding convolutional networks. In *Computer Vision — ECCV 2014: 13th European Conference, Zurich, Switzerland, September 6–12, 2014, Proceedings, Part I 13*. Springer, pp. 818–833.

Zhang, L., Rao, A., and Agrawala, M. (2023). Adding conditional control to text-to-image diffusion models. In *Proceedings of the IEEE/CVF International Conference on Computer Vision*, pp. 3836–3847.

Zhou, B., Lapedriza, A., Khosla, A., Oliva, A., and Torralba, A. (2017). Places: A 10 million image database for scene recognition. *IEEE Transactions on Pattern Analysis and Machine Intelligence*, **40**(6), 1452–1464.

Index